SCIENTISTS and TECHNOLOGISTS

by
Irene M. Franck
and
David M. Brownstone

A Volume in the Work Throughout History Series

Facts On File Publications
New York, New York • Oxford, England

Scientists and Technologists

Library of Congress Cataloging-in-Publication Data

Franck, Irene M.
Scientists and technologists.

(Work throughout history)
Bibliography: p.
Includes index.
Summary: Explores the role throughout history
of the occupations involved with science and technology,
including alchemists, biologists, computer scientists,
engineers, and physicists.
1. Scientists—Juvenile literature. 2. Technologists
—Juvenile literature. 3. Science—History—
Juvenile literature. 4. Technology—History—Juvenile literature.
[1. Scientists—History. 2. Technologists—History.
3. Science—History. 4. Technology—History.
5. Occupations—History] I. Brownstone, David M.
II. Title. III. Series: Franck, Irene M. Work
throughout history.
Q147.F74 1988 509.2'2 87-19959
ISBN 0-8160-1450-7

Printed in the United States of America

10 9 8 7 6 5 4 3 2 1

Contents

Preface

Scientists and Technologists is a book in the multivolume series, *Work Throughout History.* Work shapes the lives of all human beings; yet surprisingly little has been written about the history of the many fascinating and diverse types of occupations men and women pursue. The books in the *Work Throughout History* series explore humanity's most interesting, important and influential occupations. They explain how and why these occupations came into being in the major cultures of the world, how they evolved over the centuries, especially with changing technology, and how society's view of each occupation has changed. Throughout we focus on what it was like to do a particular kind of work—for example, to be a farmer, glassblower, midwife, banker, building contractor, actor, astrologer, or weaver—in centuries past and right up to today.

Because many occupations have been closely related to one another, we have included at the end of each article references to other overlapping occupations. In preparing this series, we have drawn on a wide range of general works on social, economic, and occupational history, including many on everyday life throughout history. We consulted far too many wide-ranging works to list them all here; but at the end of each volume is a list of suggestions for further reading, should readers want to learn more about any of the occupations included in the volume.

Many researchers and writers worked on the preparation of this series. For *Scientists and Technologists*, the primary researcher-writer was David G. Merrill. Our thanks go to him for his fine work; to our expert typists, Shirley Fenn, Nancy Fishelberg, and Mary Racette; to our most helpful editors at Facts On File, first Kate Kelly and then James Warren, and their assistant Claire Johnston and later Barbara Levine; to our excellent developmental editor, Vicki Tyler; and to our publisher, Edward Knappman, who first suggested the *Work Throughout History* series and has given us gracious support during the long years of its preparation.

We also express our special appreciation to the many librarians whose help has been indispensable in completing this work, especially to the incomparable staff of the Chappaqua Library—director Mark Hasskarl and former director Doris Lowenfels; the reference staff, including Mary Platt, Paula Peyraud, Terry Cullen, Martha Alcott, Carolyn Jones, and formerly Helen Barolini, Karen Baker, and Linda Goldstein; Jane McKean and Marcia Van Fleet and the whole circulation staff—and the many other librarians who, through the Interlibrary Loan network, have provided us with the research tools so vital to our work.

Irene M. Franck
David M. Brownstone

Introduction

Scientists and *technologists* are the people primarily responsible for what we generally call "progress." These experts are often little understood by the general population because much of their work is intellectual and seems "academic." It is only when their work is translated into world-shattering, life-changing practical uses that most people understand the power of science and technology. In truth, even the most theoretical of sciences generally has had a very solid, practical base in its early history.

Mathematics, the "queen of sciences," is probably the least understood of all the scientific disciplines. But it originally sprang from very practical needs to count and measure things. The study of geometry, for example, has its roots in the necessity to be able to clearly mark out plots of land, as modern *surveyors* do. *Mathematicians* looked at the practical world around them and saw underlying patterns of shape and motion, which they expressed through the language of numbers. That basic approach still underlies even the most abstract of modern mathematics. Every other science is, to a greater or lesser extent, dependent on the general, yet precisely formulated principles of mathematics. One particular specialty within mathematics is statistics, the science of probability. *Statisticians* have employed their special skills in areas as diverse as gambling, political polling, musical synthesis, and space exploration.

Perhaps the science most closely related to mathematics is physics. Once the word *physicist* meant simply *scientist*. But in modern times, as the sciences divided into specialties, the physicist came to be a scientist who focused on questions of force and motion. Great physicists from Newton to Einstein and beyond learned to express the patterns of structure and movement in the universe through simple, yet powerful "laws," expressed as mathematical formulas. The results of their work are all around us.

Mathematics was also closely associated with the need to calculate time and the seasons, through devices such as calendars, and to predict future events, such as eclipses, from observation of the heavens. These activities gave rise to the related disciplines of astronomy and astrology. *Astrologers* attempt to predict the course of human events from the pattern of the stars in the skies at a given time, especially at a person's birth. In most (but not all) parts of the world today astrologers are seen as pseudo-scientists, or outright fakers. But their discipline, though discredited by modern science, is far older than its more respectable cousin, astronomy. Only in relatively modern times, especially since the 16th century, have *astronomers* developed the "harder" science of studying the universe, using scientific methods and relying heavily on mathematics and physics.

Modern scientific chemistry has a similar relation to alchemy, a discipline now long since abandoned. From very early times, in both the Near East and Far East, *alchemists* worked to try to change "common" or "base" materials into precious metals such as gold and silver. A great deal of mystical and religious belief surrounded their work. Even so, alchemists were practical workers, who developed the basic laboratory techniques used in modern sciences. In Renaissance times, alchemy fell into disrepute, and the mysticism associated with it was gradually stripped away. What emerged from the ashes was modern scientific chemistry. In the last few centuries, *chemists* have made astonishing strides in understanding the elements that make up the universe and in putting those elements together in extraordinary new ways.

In modern times, physics has come to be so closely connected with chemistry that it is hard to tell where one ends and the other begins. Today the scientist exploring the tiniest particles of matter could be either. Where Galileo and Newton—and even Einstein—could work and think very much on their own, making general observations with relatively crude instruments and formulating great laws about how the universe operates, today's physicists are often found in huge laboratories, where massive machines help them explore the makeup of atoms in the pursuit of greater knowledge of the universe.

Mathematics and the laws of traditional physics have also much influenced the work of *engineers*. Whether in building roads and great structures, developing machines of war, or creating the countless devices that make our lives easier, it is engineers who put into practical use much of what has been learned in the other more theoretical or experimental sciences. Engineers draw not only on physics and mathematics but also in modern times very much on chemistry, as chemists have developed extraordinary new compounds to serve the many purposes of modern life.

In the last few centuries, a special group of technicians—*scientific instrument makers*—have worked with scientists to create devices such as the microscope and the telescope, which made much of the modern

Scientific Revolution possible. In the 20th century, *computer scientists* have combined mathematical and engineering expertise to create *computers*, the new, astonishingly powerful instruments that bid fair to revolutionize our lives.

Some sciences have not relied so heavily on mathematics. Biology, for example, was for most of its history a descriptive science, devoted to detailing and categorizing the anatomy, physiology, and life patterns of living things. And such activities still occupy the main working lives of many *biologists*. But in modern times biology has moved ever closer to chemistry, with the two practically merging in biochemistry, or organic chemistry, which focuses on the complex chemical makeup of molecules found in living matter.

Geology, the study of the structure of the Earth, was also largely a descriptive science in its infancy. But gradually it has drawn close to other sciences. As *geologists* have come more and more to understand the makeup of the Earth, they have drawn on knowledge developed by physicists, chemists, biologists, engineers, computer scientists, and statisticians to help them decipher the extraordinary history of the physical Earth from samples and observations.

While geologists study the physical Earth, *geographers* study the relationships between human beings and the various regions of the globe they inhabit. Though their discipline, too, was largely descriptive in early modern times, they are drawing more and more on biological studies and statistical sampling techniques to give them different views of how humans interact with the environment around them. Drawing on information from geographers and others, *cartographers* draw the maps that have been so essential in this modern age of travel.

For most of history, these many scientific disciplines did not exist as distinct specialties. Only around the 17th century, with the onset of the Scientific Revolution, did these various specialties and their sub-specialties break off from one another. In modern times, however, with increasing knowledge, the sciences seem to be drawing together once again, as it becomes clearer that scientific knowledge is not divided up into neat little parcels, but is one great spectrum of learning.

Alchemists

Alchemists were early scientists who tried, or sometimes claimed, to transmute "baser metals" into gold. As such, they are viewed by people today with a mixture of suspicion and mild amusement. In fact, although alchemists operated under mistaken assumptions, they carried on a wide range of activites and made considerable advances in practical laboratory work. These laid the basis for the laboratory techniques used in modern science and technology, especially in chemistry, metallurgy, and pharmacology. For over 2,000 years alchemy engaged the attention and activity of many *laboratory workers, metallurgists*, and *clerics* in several parts of Eurasia, including Egypt, China, the Near East, the Islamic world, and Europe. Like scientists today, many alchemists worked unsung, but others were highly regarded and well-known in their time.

Alchemy's roots are not precisely known. That it was practiced more than 2,000 years ago is indicated by a Chinese law of 175 B.C. that prohibits the practice of creating false gold by alchemy. An Assyrian tablet dated to the seventh century B.C. is thought by its discoverer to refer to the transmutation of silver into gold. The Chinese alchemist Dzou Yen is known to have worked in the fourth century A.D., and the Alexandrian alchemist Zosimus wrote in 300 A.D. of the then-ancient temple of Ptah, god of *gold smelters* and *goldsmiths*, at Memphis, in Egypt.

Alchemy flowered for the first time in the early centuries of the Christian era, in the Near Eastern city of Alexandria, Egypt, one of the few cosmopolitan cities of its time. Near Eastern alchemy, following Greek scientific theory, assumed that all substances could be transformed into other substances. (The idea that substances such as gold and silver were distinct, unchangeable elements did not emerge until the late 18th century.) Alchemists focused their main attention on the transmutation of other, "baser" metals into gold, which was thought to be the most perfect of substances.

Some of the most effective practical scientists in Alexandria were the Greek-speaking Egyptian and Jewish alchemists. These included Zosimus, who wrote an encyclopedia of alchemy in about 300 A.D., and his sister, Theosebeia, who also practiced alchemy. Another highly regarded practical scientist, Mary the Jewess, pioneered in several areas, especially in distillation techniques. These and many others made Alexandria a center of practical scientific development in metallurgy and laboratory work. They were particularly skillful in working with the nonferrous metals of the Bronze Age—gold, silver, copper, lead—and the dyes that could be developed in the course of working with these materials. Theirs was the most pragmatic, most enduring occupational line in alchemy, leading directly to the modern laboratory. Whatever beliefs they held about other aspects of their lives, these were practical metallurgical and chemical workers, who contributed substantially to the development of modern science.

Chinese alchemy, developing at the same time or perhaps even earlier, focused both on the transmutation of metals into gold and on the development of an elixir of immortality. The latter was an idea apparently absent from Near Eastern alchemy. Early Chinese alchemists were *scholars* and*scientists*, such as Wei Po-Yang, author of an early alchemical book in 142 A.D., and Liu Hsiang, a scholar attached to the court of a Chinese emperor as an alchemist in 60 B.C. Liu Hsiang's charge was to make alchemical gold, not as a store of value, but as a step toward immortality. For the Chinese the making hf gold was but a step on the way toward achievement of an elixir of life. Gold was thought to be imperishable, and to eat from vessels of gold was thought to be one road toward immortality.

No substantial links have yet been found between Chinese and Near Eastern alchemy, although modern archaeology continues to find more and more relationships between the major early civilizations on both ends of Eurasia. What does seem clear is the striking similarity between the objects of scientific inquiry in China and the Near East during the period of the early alchemists. Chinese alchemy did develop the concept of a catalytic substance, which would be the prime mover in transforming baser metals into gold. This concept later appeared in the Islamic world and Europe as the *Philosopher's Stone*, an idea that inspired never-ending search, endless speculation, and mystical discussion in the mass of alchemical literature that surrounded alchemy in medieval European culture.

By the sixth century A.D. the practice of alchemy had declined in both China and Egypt. By then, however, the Nestorians, the ancient Christian religious denomination of Persia, had carried Greek science, and with it alchemy, into the Arab world. There alchemists flourished for nearly a thousand years. The religion of Islam arose in the Arab world in the seventh century, and the Arab

Moslems set about creating an empire that ultimately stretched from Spain to India. From about 750 A.D., the Abassid dynasty consolidated the Islamic empire, which was, with China, one of the two great cultures of the time. Under the Moslems, alchemy was a major science. Moslem alchemists, in a culture that merged secular and religious power, were often attached to the *clergy* and *nobility*. But they were still *chemists* who built on the work of both the Greek and Chinese alchemists before them. By the end of the 12th century, they commanded a considerable body of practical chemical knowledge and effective laboratory technique. The most famous of all Arab chemist-alchemists was Abu Musa Jabir, known to later Europeans as Geber, who worked in the eighth century. Another well-known Moslem alchemist was the ninth-century Persian physician and encyclopedist, Zakariya al-Razi.

Moslem alchemical works began to be translated into Latin, mainly by Jewish scholars, in the 12th and 13th centuries, as Europe began to develop the intellectual life that would lead to the Renaissance and to the Scientific and Industrial Revolutions. By the dawn of the 13th century, alchemy was beginning to sweep across Europe.

European alchemists were, at the start, usually to be found in the clergy, that refuge of all the learned professions in pre-Renaissance Europe. But in the 13th,

In workplaces like this one at Pompeii, alchemists developed basic laboratory techniques. (From Museum of Antiquity, *by L.W. Yaggy and T.L. Haines, 1882)*

Using the furnace at the center of the shop, alchemists experimented with early chemical analysis. (University of Prague)

14th, and 15th centuries, more and more Europeans outside the clergy sought to transmute baser metals into gold, pursued the Philosopher's Stone to make that transmutation possible, and created an enormous body of mystical writing related to alchemy. And for those three centuries, an increasingly large body of charlatans preyed upon the gullible with false claims of successful transmutation—the only gold produced being that which passed from the gullible to the fraudulent. It was in this period that alchemy began to be viewed in Europe as mere mysticism tinged with fraud. But alchemy, as the great early scientist and clergyman Roger Bacon saw in the 13th century, was two-sided, carrying a load of mysticism upon a solid base of developing science.

As the ideas of the Scientific Revolution developed, the

ideas of alchemy fell into disrepute. At the same time, however, the laboratory methods of the alchemists were being adopted and developed by the chemists and other laboratory practitioners who followed them. This process, started in the 16th and 17th centuries by such rationalists and scientists as Galileo and René Descartes, was finished by such scientists as Antoine Laurent Lavoisier, the 18th-century French scientist who was one of the pioneers of modern chemistry. In our time, except for isolated cults and individuals, the occupation of alchemist is no more.

For related occupations in this volume, *Scientists and Technologists* see the following:
Chemists

For related occupations in other volumes of the series, see the following:
in *Artists and Artisans*:
Jewelers
in *Healers* (forthcoming):
Nurses
Pharmacists
Physicians and Surgeons
in *Manufacturers and Miners* (forthcoming):
Metalsmiths
in *Scholars and Priests* (forthcoming):
Monks and Nuns
Priests
Scholars

Astrologers

Astrologers are people who track the patterns of stars and other celestial bodies in an attempt to decipher some divine will at work in the universe—a fate they believe will ultimately have profound effects on the personal life of each individual. Today Westerners frequently think of astrology as a superstition and astrologers as quasi-scientists at best. In ancient times, however, astrology was first among the exact sciences, rooted as it was in mathematical calculation. The earliest astrologers were among the greatest scientists and intellects of their time, a situation that lasted for two to three thousand years right into early modern times.

Astrology took on two forms in ancient times: mundane and divinational. *Mundane astrology* was akin to, and eventually developed into, modern *astromony*; it was a rather objective and curious inspection of celestial bodies, systems, and events. *Divinational astrology*, the more dominant form for thousands of years, was based on observations of the regularity and periodicity of the movements of the sun, moon, planets, and stars. Its particular concern was in discerning from such movements the will of the gods and the fate of individuals.

The peoples of Mesopotamia were the most skilled in the practice of "divination by the stars." The Chaldeans and Babylonians, in particular, came to be renowned for their abilities and knowledge in the field. Agricultural concerns probably first encouraged them to research the course of extraterrestrial events and motions. They were directly affected—as peoples of all premodern societies were—by the changes of seasons, natural growth cycles, and events such as earthquakes, volcanic eruptions, rain, and wind. Without a proper balance of all these things, they faced starvation, disease, and destruction by natural forces. It was in this atmosphere that Mesopotamian astrologers began to seek ways of discovering and plotting the courses of the heavenly bodies, which were generally regarded as possessing great influence over nature. They eventually attached elaborate mythologies and cosmologies (theories about the structure of the universe) to their findings and speculations.

Ancient *magicians* and astrologers both practiced divination by plotting celestial patterns and interpreting heavenly events. Eventually it came to be assumed that these professionals occupied themselves with charting stars that could both indicate and cause terrestrial events, and they came to be retained and thereby employed chiefly by royal households. Royal astrologers helped *kings* and *leaders* of state to foresee coming events and to interpret omens regarding crucial decisions. For instance, astrologers might interpret a fallen star or partial eclipse as an omen that should be acted upon. The specific nature of such action was determined by those in power, often following close consultation with the royal astrologer or a team of astrologers. Not suprisingly, astrologers were frequently at odds with each other in their interpretations of celestial occurrences—and hence in their advice.

As a group, ancient astrologers had rather few opportunities for strictly professional development. While royal astrologers were highly regarded, employment was available for only a few—and even these were rarely secure in their jobs. Losses in war or natural calamities were frequent causes for dismissal of court astrologers who had failed to warn or properly advise their lords on such matters. Most astrologers came from the priestly class, where they received their training in understanding the will and action of the gods through celestial observations and calculations. And most lived quite simply and far from public recognition in the temple complex, where they quietly studied and recorded the happenings and formations in the heavens. Even the temple astrologers, however, were frequently consulted by royal families, particularly in times of trouble. The less respected of them shared other priestly duties, often spending only a small amount of their time on astrological speculation or recording.

Ancient astrologers commonly believed that extraterrestrial bodies were not natural things but stellar configurations of gods, which made seemingly willful movements throughout the heavens. Eventually, it was recognized that such movements were not so capricious. Definite patterns, regularities, and periodicities came to be recognized by the more scientifically oriented astrologers. By the first thousand years B.C. there had developed a general knowledge of the Sun's annual course, the various phases of the Moon, and the periodic movements of the planets. Several cultures had developed useful solar and lunar calendars.

Egyptian astrologers followed the lead of the Chaldeans and Babylonians in the mathematical computation of such patterns. They charted the risings (every 10 days) of 36 *decan stars*—one for each 10 degrees in the 360-degree circle that represented the heavens. The whole circle, called the *zodaic*, was later divided into 12 equal parts called *signs*, each of which was named for a particular constellation. The decan stars were later translated by the Greeks into the *horoskopos*, the foundation of personal astrology. In these ancient times, even relatively scientific information concerning the heavens became part of the magico-mythological system employed by astrologers.

The Mesopotamian forms of astrology spread to many parts of the world, most importantly India and China. But it was the Greeks of the Classical Age who really developed the art of divinational and personal astrology. The Greeks identified stars with specific gods. Borrowing the Mesopotamian advances in mathematics, the Greeks added another dimension in the calculation of extraterrestrial orbits, through their development of spherical geometry. Greek astrologers reconciled this knowledge with Greek mythology, developing the *zodiac*, the 12 stellar signs of the year, each presided over by a specific spirit. They extended the royal art to pertain to all humans; the result was the new *genethlialogical*, or personal, astrology. In the second century B.C. the Greeks produced the first textbook on astrology in Alexandria, Egypt. In essence, the Greeks transformed the older forms of mundane astrology into elaborate forms of personal and divinational astrology, which went far beyond the earlier limits of the science.

Most Greek intellectuals supported the work of the astrologers. Plato and the Pythagoreans—even Hipparchus, one of the earliest astronomers—found considerable merit in such attempts to understand and interpret the will of the gods through celestial manifestations. The Stoics taught that there was an undeniable *sympathy* between what caused the motions in the heavens and what caused the destiny of each individual on Earth, which was assumed to be the center of the universe.

Many Romans, on the other hand, were fearful of the strong influence that astrologers (*mathematici*) had on politics and society. Some emperors even banished those

The Moslem astrologers on right and left are casting the horoscope for their prince's newborn son. (From Magamat, *by al-Hariri, 13th century, Bibliotheque Nationale, Paris)*

who performed the royal art. Augustus and several of the later emperors, however, welcomed the advice of the royal astrologers whom they sponsored.

The growth of Christianity did not spell the waning of the profession, as one might expect. Christianity is based on the doctrines of free will and the divinity of a single God. Astrology presupposed the doctrine of determinism as seen in the divinity of the stars. Nonetheless, the two systems were integrated for a time, probably because of what was seen as their mutual dependence on one another. Christian astrologers saw the work of the creator in stellar movements, and continued their research in the field from that perspective.

With the fall of the Roman Empire, the Greek form of the royal art was nurtured and developed primarily by the Arabs. With the eventual influence of Islamic science on the medieval world, royal astrologers became important once again, enjoying the luxuries of the great European courts, where they peddled their "informed" advice. In an age when men were forbidden to attend at the birth process, often on pain of death, the only exception was the astrologer, who was present to record the precise instant of birth, for the most accurate horoscope.

In the 13th century both Dante and St. Thomas Aquinas integrated astrologers' ideas of causation into their own philosophies. University *professors* found astrology a good subject for attracting students—no small point in an age when their income depended on the size of their classes. The 14th-century universities at Paris, Bologna, and elsewhere sponsored chairs of astrology, typically held by prominent astrologers of the day. As late as the Reformation we find well-informed, scientifically oriented men believing steadfastly in the merits and skill of the astrologer. Even leaders in the expanding science of astronomy—notably Tycho Brahe and Johannes Kepler—were among the admirers of the royal art.

Astrologers lost their reputations as scientists largely after the 16th century, when Nicolaus Copernicus showed that the Earth revolved around the Sun, rather than the other way around. This changed Western thinking on the structure of the universe. New evidence from the growing science of astronomy continually eroded the cosmological systems that Greek astrology had been founded upon. More and more, people who were interested in studying the heavens became *astronomers*, while those concerned with God's will became *theologians*. Astrology became thought of as an outdated attempt to understand what could better be understood through rational endeavors than by superstition and magic. Very few intellectuals retained interest in the field.

Today, astrologers are generally regarded as eccentric and superstitious in an age dominated by science and rationalism. Nonetheless, the profession lingers on in the West. Many contemporary astrologers are merely *entertainers* who interest people with their personal horoscopes and predictions of future events both personal and universal. More serious forms of the art are carried on in astrological societies and associations, which insist that the will of the cosmos, or God, or of the

Even in the science-oriented 20th century people advertise "authentic astrology." (By Russell Lee, Library of Congress, in Lousiana, 1938)

spirits can still be detected through stellar observation, as much today as ever before. And in those parts of the world where everyday life has been relatively untouched by the ideas of the Scientific Revolution—in India, for example, and many parts of Africa—astrologers are still numerous and widely consulted.

For related occupations in this volume, *Scientists and Technologists*, see the following:

Astronomers
Mathematicians

For related occupations in other volumes of the series, see the following:

in *Healers* (forthcoming):
Midwives and Obstetricians
in *Leaders and Lawyers*:
Political Leaders
in *Scholars and Priests* (forthcoming):
Priests
Scholars

Astronomers

Astronomers are scientists who try to learn about celestial events and occurrences through observation and calculation. All early people have been drawn to record and systematize information about the heavens. Living on the land, they found it essential to keep track of the seasons. Gradually they developed calendars, and later they found ways to predict such unusual events as lunar and solar eclipses. Peoples all around the world have left evidence of their efforts to record such information in useful ways. Some of the earliest marks made by prehistoric humans, on bones found in caves occupied tens of thousands of years ago, may have been calendrical records. Later peoples, even into modern times, went to extraordinary lengths to mark the seasons as they passed. Such *paleoastronomers* created massive outdoor observatories in the process, of which Stonehenge is probably an example. In these early times, astronomical lore, like all learning, was probably the province of *priests*.

As humans entered historical times, astronomy was still linked with religion—not surprisingly, since theories of astronomy were, in fact, theories of the universe. With prediction of the future still a prime desire, *astrologers* were important figures in these early communities, and it was they who carried out most of the astronomical research. They were generally interested in the heavens less out of scientific wonder than a desire to support religious or mystical theories. The Chaldeans developed a *geocentric* theory, which presupposed that the Earth was the center of the universe. The universe was conceived of as a sort of enclosed dome from which stars hung like light bulbs. Later, the Greeks argued about this idea. The Pythagoreans posed a new *heliocentric* theory, which stated that the Earth and planets all revolve around a central sun. Aristarchus of Alexandria, one of the earliest astronomers, developed this theory more fully. Aristotle, though, preferred the geocentric model, since it focused on humankind as the most important beings in the universe.

Hipparchus and Ptolemy, the most influential of the ancient Western astronomers, supported the Aristotelian view in the second century B.C. and the second century A.D., respectively. Ptolemy's summary of ancient astronomy, the *Almagest*, was strictly slanted in favor of Aristotle and Hipparchus, and away from Aristarchus, who had so eloquently described the heliocentric view. The *Almagest* would later become the most revered sourcebook of the Islamic astronomers, who saved the science from its complete collapse in the West following the fall of the Roman Empire. The Aristotelian view of a stationary earth encircled by revolving planets and celestial bodies was also the ideal one for Christian *theologians* to integrate into their religious cosmology. As a result, the Ptolemaic model of the cosmos lasted until 1600 A.D., when the heliocentric theory was finally determined to be the more accurate of the two.

Ptolemy, with the muse Astronomy looking over his shoulder, is using the tools of his science, the quadrant and armillary sphere. (From Margarita Philosophica, *Gregor Reisch, 1508)*

Specialist astronomers also emerged elsewhere in the world, most notably in China. Chinese astronomy has been called the "secret science of priest-kings." From very early times, an astronomical observatory was a traditional part of the *emperor's* home. The Chinese were far more interested in astronomical observations and calculations, especially for drawing up calendars, than they were in the grand theories of the universe that entranced their Western counterparts. The emperor's formal announcement of the coming year's calendar was a major event in China. The Chinese even developed early, elaborate kinds of clocks, primarily designed to assist them in calendrical computations. Astronomers were *government officials*, given the highest status, like other *scholars* in the realm. This would be true for centuries; in the late 19th century, Viennese writer Franz Kuhnert noted with tongue in cheek:

> Probably another reason why many Europeans consider the Chinese such barbarians is on account of the support they give to their Astronomers—people regarded by our cultivated Western mortals as completely useless. Yet there they rank with Heads of Departments and Secretaries of State. What frightful barbarism!

Chinese astronomers were strikingly skilled observers, fortunately, for the strongly observation-oriented astronomical tradition of the early Near East had gone into decline. As a result the Chinese provided the only accurate records of astronomical observations for many centuries, from before the birth of Christ to a thousand years afterwards. Indeed, modern astronomers have profitably consulted these early Chinese records to confirm predictions of a comet's periodic return or the likely dating of a stellar explosion, called a *supernova*.

Like other agricultural economies, India also found astronomy important. In early Vedic lore, astronomy focused on preparing calendars of the seasons, especially for setting the time of periodic sacrifices to the gods. Astronomy on the sub-continent was closely related to astrology, a connection that has survived into modern times in India. Indian *sailors* navigated by the stars, and even *caravan leaders* studied the heavens so they could guide their caravans by the stars when forced to travel at night. India drew to some extent on Near Eastern and Greek learning, making contributions of its own, especially in the area of mathematical calculations.

The Islamic world that arose in the Near East in the seventh century drew on these various traditions. Among the many Western texts that the Moslems gathered, translated, and transcribed were astronomical texts. Ptolemy's *Almagest* had pride of place and was enormously influential, both in its theory and in its injunction to carry out astronomical observations. Groups of astronomers gathered around the brightest Islamic courts, refining calendars, observing eclipses, and calibrating the dating of the seasons. These astronomers

Civilizations around the world had astronomical observatories, like this one at Delhi. (From Lights and Shadows of Asiatic History, *1848)*

came from widely varied backgrounds; some of the finest among them were Jews and Indians. A 13th-century Islamic observatory, perhaps the first in the West, even included on its staff an astronomer from China.

Over the centuries, the Moslems kept alive an intellectual inquiry into the workings of the cosmos. Eventually a group of European astronomers revived astronomy as a science in Europe; they redirected its course of investigation between 1543 and 1800, a period that was truly the golden age of astronomy.

The new age of astronomy began in 1543, when the Polish astronomer Nicolas Copernicus published his theories refuting the Ptolemaic system, which had been considered the chief authority in the West for nearly 1,500 years. Somewhat earlier, a German cardinal named Nicholas of Cusa had suggested that the heliocentric view proposed by Aristarchus was a much simpler and more practical guide to use in calculating planetary positions than the long-adhered-to theory of geocentricity. Copernicus substantiated the heliocentric theory. In 1512 he had started work on a mathematical system that would better map planetary positions according to this new model. In time, his system apparently became dependent on the notion of *steller parallax,* or the changes in the positions of the stars relative to that of the

Earth. Parallax was necessary if the Earth were, as Copernicus had determined, in perpetual orbital motion around the central Sun, like all other celestial bodies.

The Copernican system was detailed and rather precisely calculated for the time, but it was bound to meet with stern opposition on several grounds. The most important objection was the religious one. For centuries, the Catholic church had made the Aristotelian philosophy and the Ptolemaic cosmology central to its teachings about humanity's position and role in the universe with respect to the Creator. That the Sun should be the center of the universe and the Earth just another orbiting planet was thought to in some way demote humans to a less central role in the universe—an unacceptable role for a creature thought to have been created in God's image.

The other objection to the Copernican system was a scientific one, based on the limited knowledge of the times. The theory of stellar parallax was rejected by most astronomers and other scientists because it was possible only if the Earth was in perpetual motion. This was considered impossible because, if such were the case, the scientists thought, people and buildings on the Earth's surface would fall off into space when the globe turned upside down. Gravity would be unknown until Newton's work over a hundred years later, and for at least half a century there were no instruments available by which the phenomenon of stellar parallax could be directly detected.

Copernicus was a scholarly gentleman, who was raised by an uncle who was a prince and a bishop. Like other astronomers and most scientists of the period, he was a member of the ruling class and had great love and respect for the Catholic church. He was an unlikely revolutionary and, in fact, his entire system was an elaboration of the earlier-stated suspicion of another churchman, the Roman Catholic cardinal Nicholas of Cusa. Copernicus was so concerned that his work would be ill-received by the church that, although his findings were explicitly written in a book, he delayed its publication until just before his death. Even then he dedicated it to Pope Paul III. Moreover, he personally saw no reason why his theories should be contrary to Christian theology, for they only showed that God was perhaps wiser than even the church fathers had known. He once stated this notion in the following way:

> At rest in the middle of everything is the sun. For in this most beautiful temple, who would put this lamp in another or better position than from which it can illuminate the whole thing at the same time? Thus, indeed, as though seated on a royal throne, the sun governs the family of planets revolving around it.

Copernicus was not alive to see how his book was received by the church or by other scientists, but his worst fears concerning the matter would have been confirmed if he had lived. He may be reverently thought of today as the father of modern astronomy, but at the time his views were considered by many to be extreme blasphemy. He himself was seen as a renegade, revolutionary, and heretic. Yet his thesis was broken gently to the church. Not only did Copernicus carefully dedicate his book to the Pope, he also left its publication to a Lutheran minister named Osiander. Osiander, knowing of Martin Luther's disdain for the Copernican theory, wrote an introduction that made the book seem little more than a computational exercise by its author. Moveover, according to Osiander, the author was acutely aware that no real knowledge of the universe could be forthcoming except through divine revelation. He was merely making whatever insignificant observations he could in his earthly occupation as an astronomer. Therefore, Osiander wrote:

> . . . the author of this work has done nothing blameworthy. For it is the duty of an astronomer to compose the history of the celestial motions through careful and skillful observation. Then . . . he must

> conceive and devise, since he cannot in any way attain to the true causes, such hypotheses as, being assumed, enable the motions to be calculated correctly from the principles of geometry, for the future as well as for the past. The present author has performed both these duties excellently . . . .
>
> It is quite clear that the causes of the apparent unequal motions are completely and simply unknown to this art. And if any causes are devised by the imagination, as indeed very many are, they are not put forward to convince anyone that they are true, but merely to provide a correct basis for calculation. Now when from time to time there are offered for one and the same motion different hypotheses . . . the astronomer will accept above all others the one which is the easiest to grasp. The philosopher will perhaps seek the semblance of the truth. But neither of them will understand or state anything certain, unless it has been divinely revealed to him . . . .
>
> . . . let no one expect anything certain from astronomy, which cannot furnish it, lest he accept as the truth ideas conceived for another purpose, and depart from this study a greater fool than when he entered it. Farewell.

In truth, the significance of the Copernican theory went far beyond the particular views of the work. It opened a new age of astronomy, one in which people could rely on their own reasoning abilities to actively seek to uncover the secrets of the heavens, rather than passively accept church doctrine on the subject. It was an age in which measurement and calculation became increasingly relied upon—and, in turn, became increasingly precise and reliable. Eventually, instruments were developed to aid in observation of the heavens as well as in calculation. Copernicus himself had a deep confidence in the power of human rationality; because he viewed the universe in mechanistic terms, he believed that people could unlock all the celestial secrets with their intellects.

The revolution in scientific thought introduced by Copernicus had a vast impact on all the scientific professions. Astronomers in particular became much more likely to view the world in mechanistic terms. They began spending as much time and effort making mathematical calculations about the workings of the *grand machine*, the universe, as they did actually observing it. Astronomers of this era, to no small extent, paved the way for scientists in other professions to deal with their specific investigations through mechanistic models and with confidence in their own rational capacities. But although such views may eventually have led other astronomers away from the intellectual grasp of the church (and its handmaiden, the university), their authors were hardly upstarts who sought to undermine the church's authority. Copernicus was a dedicated churchman and so, too, were other leaders of the new astronomy, notably Tycho Brahe and Johannes Kepler.

Most astronomers were still obliged to be adept in astrology and to use it as an explanation of Christianity and its teachings. Professors of astronomy found astrology very attractive to young students—no small consideration during the Reformation, when university enrollments in Europe dropped sharply. Professors were often paid according to the number of students enrolled in their courses, so astrology courses were important to their livelihoods. Even astronomers who enjoyed royal or noble patronage were frequently called upon to cast horoscopes and advise their patrons in personal and political matters. Their pay for this service was much higher than for scientific findings. Most astronomers considered their work in astrology and religion as a necessary but otherwise undesirable aspect of their jobs, although Kepler was known to have been prouder of his abilities in the mystical realms than of his critical achievements in the dynamic calculation of celestial masses.

Tycho Brahe, a Danish noble, was one of the last of the great naked-eye astronomers. As a young man he was the first to show (through actual parallax measurements) that the Aristotelian view of a fixed and unchanging universe was just not accurate. Nonetheless, he rejected Copernicus's suspicion that stellar parallax was caused by the movement of the earth. In fact, he was the last prominent astronomer to hold tenaciously to the geocentric Ptolemaic model of the universe. The king of Denmark, Frederick II, became Tycho's patron to prevent his emigration to Germany, where the most active astronomical research was being carried on. With Frederick's patronage, the Danish scholar built Europe's first astronomical observatory on Hven Island. The project was completed in 1580 at a cost of some million and a half dollars (today's value) and included a spectacular spherical globe of the heavens, five feet in diameter. Though working before the age of telescopes, Brahe was the first to properly adjust calculations to account for atmospheric refraction, the bending of light rays as they pass through the atmosphere. Brahe thought he had explained the changes in the apparent position of heavenly bodies.

Tycho Brahe was a typical scientist of this time. He was a snobbish nobleman who thought it below his dignity to

Astronomy and astrology were long intertwined in the popular mind, and the astronomer was asked to predict events by the stars. (By Jost Amman, from The Book of Trades, *late 16th century)*

work for a living or write books detailing his observations (although he did both). He placed himself among luxurious and costly instruments (most of which he designed himself) and surroundings; he even arrayed himself in court dress to engage in observational work. After Frederick II died in 1588, his successor quickly tired of Brahe's pompous and expensive habits; he refused further subsidies, so Brahe finally left for Germany. There, in his waning years, he found a new sponsor in Emperor Rudolf II, and a new student in young Johannes Kepler.

Kepler was more skilled in calculation than in observation. In fact, he used Brahe's observations, which he inherited from his master, rather than make his own. Kepler's calculations made it apparent that the future of astronomy would be dependent at least as much upon mathematics and physics as it was on direct observation. His three landmark laws of planetary motion were fully formulated by 1618, just a decade after his contemporary, Galileo, began using a new observational instrument—the *telescope*. The first two laws that he established showed that planetary orbits were actually elliptical (oval) rather than perfectly circular, another blow to the Ptolemaic and Christian views concerning the boundless perfection—which the circle supposedly symbolized—of God's universe. Then, once elliptical orbits were accepted, it followed that planets must travel at non-uniform rates around the Sun. Although Kepler, like his contemporaries, was puzzled over what force maintained the planets in their elliptical orbits, he did lend insight toward the calculation of distances between the heavenly bodies through his third law—the *harmonious law.*

At about the same time that Kepler was grounding astronomy in the physical laws of the universe, Galileo was making new observations with his *optic tube* (telescope). His observations would eventually have the effect of popularizing the Copernican theory once and for all, although the validation of its mathematical and physical basis would await the work of Newton. Galileo finally dashed all notions of perfection in the heavens with his observations of such "blemishes" as sunspots and moon craters. In the process, he pulled astronomy out of the grasp of the church and the church-oriented universities. Galileo's career and the church's reaction to his research and observational work pointed to an inherent conflict of interest in the church's involvement in academia.

Nowhere was this conflict more evident than in Galileo's call for astronomical research and theory to be based on the *scientific method.* According to this process, evidence was to be observed and assimilated into a general body of scientific information in an objective manner, without preconditions that would dictate, or at least prejudice, the results and their interpretations. By contrast, the church's notion of scientific knowledge was limited to those things found to be true and at the same time in support of church doctrine, which was unquestionably accepted as divinely revealed and therefore unalterable. At best, the church could accept Osiander's earlier contention that astronomy was a set of mathematical contrivances designed "merely to provide a correct basis for calculation." It is not surprising, given this intellectual atmosphere, that John Milton wrote in *Paradise Lost* of God's supposed view of astronomers:

> He his fabric of the Heavens
> Hath left to their disputes, perhaps to move
> His laughter at their quaint opinions wide
> Hereafter, when they come to model Heaven
> And calculate the stars, how they will wield
> The mighty frame, how build, unbuild, contrive
> To save appearances, how gird the sphere
> With centric and eccentric scribbled o'er,
> Cycle and epicycle, orb in orb.

As important as any actual observations that Galileo made was the fact that he championed the scientific method in general and astronomy as a science in particular. Like all other astronomers of the age, he was by no means above casting horoscopes. But he separated the mystical, religious, and philosophical dimensions of astronomy from its objective observations and calculations. He then determined to do what few men in any walk of life dared to do at that time: He defiantly challenged the church after 1616, when it declared belief in the Copernican system to be a heresy. His defiance was no light decision. In 1600 Giordano Bruno was burned at the stake after being charged by the Venetian Inquisition with heresy for preaching such doctrines as the infinity of space and the motion of the Earth. Nor was the Catholic church the only religious institution in the dark regarding scientific astronomy. Even in Protestant America, Harvard University—founded in 1636, four years after Galileo published his laudable defense of the Copernican system, *Dialogue on the Two Chief World Systems* —remained firmly committed to the Ptolemaic view.

To carry out his unpopular work, Galileo even disassociated himself from the church-controlled university system, in which he had first begun medical training at his father's bidding. (*Physicians* made far more money than *mathematicians* or astronomers.) He worked, instead, in close association with one of the new scientific academies, *Academia Dei Lincei,* which was established specifically to overcome the intellectual restraints enforced by the university. Galileo finally recanted his support of the Copernican system, but only while on trial—and to avoid Bruno's earlier fate. Following the trial he supposedly whispered under his breath "*Eppur si muove*!"—"and yet it [the Earth] moves." But Galileo's greatest work had been done. He had effectively demonstrated that the universe could be dealt with in mechanistic terms through calculation and rationalism. This soon became the foundation of the new astronomy, as astronomers increasingly rejected blanket church authority and sought, instead, to discover the secrets of the heavens by discerning the matematical laws of nature.

Most of Galileo's work was based on direct observation

Early 16th-century astronomers like this one would change the face of their science with new observations and theories. (Attributed to Albrecht Dürer)

using the *telescope*, a new invention that he popularized. The stage was set for a search for new and constantly improving instruments to aid astronomers in their scientific quest. However, for all the enlightenment that he offered, Galileo was unable to explain the actual motion of the Earth and other planets which the Copernican system presumed. On the basis of several experiments with falling and rolling objects, he suggested that once in motion, planets may need no further force to propel them. Following the Aristotelian view that God was the perpetual mover, medieval theologians had gone as far as to postulate that a corps of angels was employed by

God in the menial task of keeping the planets on the move.

The question of what caused an object to keep moving, once started, was an extremely puzzling one. In the 14th century, a Parisian philosopher named Jean Buridan theorized that the force that kept an object moving must be something within the object itself. He called this internal force *impetus*. Buridan suggested that *impetus* was the original, divine cause of celestial motion. Giordano Bruno and an Englishman named Thomas Digger had already demonstrated that this *perpetual motion* could be along a Euclidean straight line (such as the course of a fallen object) if there were no further interference. But interrupted patterns, which were natural in the heavens, still needed explanation.

The notion of perpetual motion was clearly one that needed to be explored, unpopular as it was with the church. Meanwhile, the observers of the heavens, now aided by telescopes, were probing another previously blasphemous idea—that of an infinite universe. But none of these ideas could progress without more concrete knowledge of the motions of celestial bodies and how they affected each other. In other words, astronomy came to focus on questions of physics and dynamics, the hard scientific knowledge needed to back up the new ideas. It was the efforts of Copernicus, Brahe, Galileo, and other pioneers in the field that made it possible for the new age of astronomers to finally remove the shackles of religion, philosophy, superstition, and eventully even astrology.

The Copernican system was finally tied together by the mathematical discoveries of England's Sir Isaac Newton, one of the greatest scientists ever to have lived. Newton was not strictly an astronomer, but many of his theories concerning dynamics (laws of motion) were worked out with reference or at least application to the cosmos. His work had profound effect on astronomy and may be considered the beginning of *astrophysics* as a separate field of inquiry.

The fundamental understanding that made Newton's

work so illuminating was that the physical laws of the Earth were the same as those of the heavens, not two different sets of laws for separate realms, as had been assumed. His law of *gravitation* was especially important to astronomers, not only because it finally gave mathematical validation to many of the theories of Copernicus, Galileo, and Kepler, but also because it prepared a new ground for exploration. Newton learned that Galileo and others were correct in thinking that a planet would travel in a Euclidean straight line through the universe, given an original impetus, which he called *force*, and freedom from obstruction. The problem with trying to conceive of such a pattern was that there was obstruction, in the form of the attractive force of the Sun. Newton calculated the Sun's attraction as being directly proportional to the sum of the masses of the Sun and another given planet (i.e., Earth), and inversely proportional to the square of the distance between them. In other words, mass and distance could be used to calculate the specific amount of gravitation of the Sun, Earth, or anything else.

Of course, this relatively simple model was complicated immensely by the fact that the heavens consisted not just of the Sun and Earth, but of numerous other planets and celestial bodies that had their own masses and distances from the Sun, and hence their own forces and orbits to be considered in the whole scheme of things. But Newton's work gave the means and direction for further study in this area. He also proved mathematically that Kepler had been correct in asserting that planets orbit in elliptical patterns rather than circular ones.

Newton's *Philosophia Naturalis Principia Mathematica*, published in 1687, gave *mechanists* and *deists* ammunition for their causes, and certainly gave astronomy great respect as an exact science. But the new astronomers were not all mechanists, who thought the universe was a great machine that just happened. Nor did all astronomers believe that new concepts of the universe—as some deists seemed to argue—were a replacement for the concept of God altogether. Newton himself had written that God's constant intervention in the cosmos was necessary, lest it be torn apart by the disorder introduced by various "perturbations" or disruptions of harmony:

> By reason of the tenacity of fluids and attrition of their parts, and the weakness of elasticity in solids, motion is much more apt to be lost than got and is always upon the decay.

Newton supposed that God restored dissipated motions through certain "active principles":

> Seeing, therefore, the variety of motion which we find in the world is always decreasing, there is a necessity of conserving and recruiting it by active principles, such as are the cause of gravity, by which planets and comets keep their motion in their orbs and bodies acquire great motion in falling, and cause of fermentation, by which the heart and blood of animals are kept in perpetual motion and heat, the inward parts of the earth are constantly warmed and in some places grow very hot, bodies burn and shine, mountains take fire, the caverns of the earth are blown up, and the sun continues violently hot and lucid and warms all things by his light. For we meet with very little motion in the world besides what is owning to these active principles. And if it were not for these principles the bodies of the earth, planets, comets, sun and all things in them would grow cold and freeze, and become inactive masses; and all putrefaction, generation, vegetation, and life would cease, and the planets and comets would not remain in their orbs.

Of course, not all scholars were completely pleased with Newton's concessions to God's power. The famous German philosopher Leibniz, for one, asked if Newton really thought God was so poor a craftman that he had to constantly be on guard to operate and repair that which He had created and put in motion. Astronomers and

Soon after optical developments made it possible, astronomers had large telescopes for long-distance rooftop viewing. (From Selenographia, *by Johannes Hevelius, 1647)*

scientists like Newton were becoming more sophisticated and probing, but not necessarily less religious. They simply had different views of the cosmos than those that had been held by the church ever since ancient times, when very little factual evidence had been available upon which to formulate accurate hypotheses.

After Newton, the history of the occupation of the astronomers is a matter of recording new findings; the foundations had been set for the direction of research. From the 18th century up to the present, astronomy has seen new developments in the measurements of distances to the stars and planets; the learning of the physical conditions of the stars through a method known as *spectroscopy*; the revelation, through development in astrophysics, of energy sources responsible for making the Sun and stars shine; the discovery of the red shift of quasars and the outer galaxies (as we will discuss later), which suggests to many contemporary astronomers that the entire universe is constantly expanding; and many other findings of significance.

In 1728 James Bradley, using a 212-foot telescope, was able to detect and explain the "aberration of light" which made it necessary to angle the telescope slightly because of the gradual annual displacement of stars in relation to the Earth. The shift in light that was beyond the expected degree of parallactic displacement proved beyond doubt Copernicus's contention that the Earth did, in fact, revolve. Bradley did not detect the actual parallax of the stars, nor could he calculate their distinct distances. These things were first accomplished in 1828 by Friedrich Bessel. It was after Bessel's findings that astronomers were first really able to look beyond the immediate solar system to more distant stars.

Working at the end of the 18th and the beginning of the 19th centuries, Pierre Simon Laplace summarized a half a century of work by Continental mathematicians using calculus (inspired by Newton) in the service of astronomy. Laplace's five-volume work, *Celestial Mechanics*, was a landmark in dynamical astronomy. Napoleon, who had appointed Laplace to the post of France's minister of interior, remarked that it seemed to cover all causes for celestial events except God. Laplace responded that "I had no need of that hypothesis"; his close associate Lagrange added, "Ah, but it is a beautiful hypothesis just the same. It explains so many things." Astronomers had generally come to the point at which they believed they could depend on their own rationality to explain the cosmos. If things like the "original cause" could be attributed to God, well and good. But that was more a matter for philosophers and theologians, not astronomers.

In 1781 William Herschel shocked the world with his discovery of a new planet, Uranus. In the 1850's, G.R. Kirchhoff, professor of physics at Heidelberg, developed a new means for deciphering the chemical properties of celestial bodies. The *spectroscope* was a revolutionary device that dissected and analyzed the light a telescope could bring in. By 1929 the American astronomer E.P Hubble had recognized that most of the galaxies beyond the Milky Way (Earth's galaxy) have spectra that are shifted toward the red end of the spectrum. He immediately associated this *redshift* with the *Doppler shift* suggested by Christian Doppler in 1842. According to this view, if spectra are found to shift toward the red end, the source of light is moving away from the observer; the exact opposite is true of spectra shifting toward the blue extreme. As a result, Hubble was able to predict that distant galaxies and bodies are constantly moving away from the Milky Way—in other words, that the universe is constantly exploding or expanding, perhaps in the aftermath of a "cosmic explosion" that occurred some 5,000 million years ago. Another interpretation of the redshift is that radiation (part of the light source given off by extraterrestrial bodies) ages, in the sense that its wavelengths increase proportionately with age. Since such wavelengths are received as redshift spectra, the galaxies farthest away—that is, those having the "oldest" radiation emissions—have the greatest redshifts in relation to Earth, even though they may be perfectly stable and stationary.

Cosmic radiations of radio wavelength were first discovered accidentally in 1931 by Karl Jansky, a radio engineer for Bell Telephone Laboratories. He determined that ionospheric radio static resulted from natural, but extraterrestrial, emissions of radio waves. Radio frequency energy was soon found to be a part of the electromagnetic spectrum that, unlike light waves, is able to penetrate the dust clouds that obscure the Milky Way's galactic center. *Radio astronomers* are modern specialists trained to interpret *microwaves* (the shortest radio waves) rather than light waves. In this way they gather information about our galaxy through radio telescopes that could never be obtained through optical telescopes or through the methods of spectroscopy.

Generally speaking, advances in instrumentation and calculation, along with the aid of computerization, have

In the 19th century astronomy became a popular amusement, with telescopes set up in parks for viewing the heavens—for a small fee. (By Stanley Fox, from Harper's Weekly, *July 25, 1868)*

made the secrets of the heavens increasingly available to industrious astronomers. Spectroscopic studies have allowed astronomers to look beyond our galaxy, and computer-bearing satellites with photographic capabilities have assisted them since the end of World War II. Cameras now do most of the observational work through photoelectric devices, so astronomers are freed to do calculations and interpretations.

Today's astronomers are, as astronomers have always been, a select group. For the men and the increasing number of women hoping to fill one of the relatively few openings available, a Ph.D. and extensive training in physics, mathematics, and computer-handling—beyond the basic astronomical subjects—are necessary. As science moved into a university setting in the last century, so did astronomy. Once largely dependent on private funds or patronage, most astronomers are now employed in colleges, although some are found in

Modern telescopes would gradually come to dwarf their users. (Authors' archives)

government positions, in the aerospace industry, or in private planetariums built for public education.

Timeless questions persist concerning such subjects as the creation of the universe. Did the universe begin with a sudden "Big Bang" or is it maintained through a "Steady State?" Are there "other universes"—as the works of Hubble and Baade and others seem to indicate?

Perhaps the most intriguing and basic of all these topics is the one concerning the source of the force that holds the universe together. Newton had insisted that God was the original cause of the force and gravity of the universe. He was attacked on this point for suggesting that a "perpetual miracle" kept the universe in motion. Newton once posed the problem in a letter to the Bishop of Worcester:

> It is inconceivable that inanimate brute matter should without the mediation of something else which is not material, operate upon and affect other matter without mutual contact, as it must be if gravitation, in the sense of Epicurus, be essential and inherent in it. That gravity should be innate, inherent, and essential to matter, so that one body may act upon another at a distance through a vacuum, without the mediation of anything else, by and through which their action and force may be conveyed from one to another, is to me so great an absurdity that I believe no man who has in philosophical matters a competent facility for thinking can ever fall into it. Gravity must be caused by an agent constantly according to certain laws, but whether this agent be material or immaterial I have left to the consideration of my readers.

Today, the "source" is thought of by some scientists in terms of radiation or "fields of force" that act as mediums for gravity. But even Albert Einstein—whose earthshaking general theory of relativity has seemingly modified Newtonian Law to the extent that it explains even the slightest factors perturbing planetary motions—suggested that this is one problem that will probably never be properly solved, through either intellect or calculation.

For related occupations in this volume, *Scientists and Technologists*, see the following:

Astrologers
Mathematicians
Physicists
Scientific Instrument Makers

For related occupations in other volumes of the series, see the following:

in *Artists and Artisans*:

Clockmakers

in *Financiers and Traders*:
Merchants and Shopkeepers
in *Helpers and Aides*:
Drivers
in *Leaders and Lawyers*:
Political Leaders
in *Scholars and Priests* (forthcoming):
Priests
Scholars
in *Warriors and Adventurers*:
Sailors

Biologists

People have always had an interest in living things. Early human beings were curious about animals and plants that could supply food, clothing, and shelter; and they needed to learn about their own bodies in order to prevent sickness, promote health, and cope with death. The earliest close observers of animal and plant life were *hunters* and *farmers*. The first to consider the human body as an entity worthy of closer study were the *healers*, whose business it was to repair the body when injured or ill, and to prepare it both physically and spiritually for death. The first real *scientists* or seekers of knowledge concerning life forms were the Greek philosophers from Athens during Classical times.

The Greeks were "real" scientists in the sense that they questioned the actual nature of life—but not in the modern, objective sense, by experimentally observing and quantifying biological data, which might later be synthesized into testable hypotheses and conclusions. They did not weigh or measure or compare phenomena (things or events) in a way that was purely experimental and as far from subjective opinion or belief as possible. In fact, they rarely even used experimental—what we would now call *scientific*—techniques in deriving general conclusions and statements concerning their subjects. But we should not be too quick to condemn the Greeks for failing to apply objective methods to what they believed were philosophical and religious inquiries.

To the Greeks, biology was a matter of reason and philosophical dialogue, like all other sciences. Empedocles, for instance, suggested that all life was made up of four basic and irreducible "roots" or elements: water, fire, earth, and air. The two forces that joined and separated them he called *harmony* and *discord*. The "inner heat" of life, he claimed, was created by the flow of blood, thus attributing to the heart the central role in the vascular system. The latter point has been proven true, of course, but Empedocles arrived at his conclusions purely through logic, not through the sort of experimentation which later became the ultimate test of scientific hypotheses.

There was, however, some degree of objective inquiry among the ancient Greeks, which was precisely what distinguished them from other early inquirers. Alcmaeon of Crotona, a physician, recognized the presence of the optic nerve and the Eustachian tube, the latter connecting the ear to the nose. These were only rediscovered in biology in modern times. He also wrote about embryonic development. The study of *anatomy* was first properly undertaken by Diogenes of Apollonia, who gave a simple description of the vascular system. Polybus supposed that four *humors* or vital fluids (blood, phlegm, black bile, and yellow bile) comprised the human body.

Perhaps the greatest understanding of biology in broad terms at that time came from Hippocrates, the famous Aegean physician from the island of Cos. It is true that he had many misconceptions concerning the workings of the human body. He thought that the heart was the seat of the intellect and that when a person drank water, it mixed with blood in the lungs. Despite his lack of accurate information (in which he was not alone), Hippocrates had clear insight concerning the general condition of the body. Many modern physicians would be fortunate to have as keen an understanding of their patients. Perhaps his greatest contributions to the study of biology were his teachings that the body possesses its own natural means for curing illness (which itself results from natural causes) and that each person possesses individual biological characteristics. On this last point, the Hippocratic writings include a simple adage that only recently has been reconsidered in its vital complexity: "One man's meat is another man's poison." Hippocrates also studied and classified animals and performed dissections on them to increase his direct knowledge of life. His teaching concerning the body's ability to cure itself of disease formed the core of biological research in the medical profession for many centuries to come.

In the fourth century B.C., Aristotle made many notable observations in his elaborate studies and classifications of animal life. The most profound was his theory of *Scala naturae*, in which Aristotle suggested that there was a natural scale of succession from the least to the most complex animals. However, this was not to suppose that there was any sort of evolution of one species into another: it was assumed that there was a rigid separation between the species. The Greeks also carried out studies in *botany* and the relation between plants and the curing of disease. Dioscorides was able to list some 600 medicinal plants in his *De Materia Medica*. While biology did not play a major role in the academic pursuits at Alexandria, Herophilus investigated both the brain and the nervous system at the famous library there, and Erasistratus was the first to differentiate the cerebellum (an important part of the brain). Both were accused of permitting public vivisections of human beings, such as convicts, for academic purposes.

In the second century A.D. the Greek physician Galen undertook a life of observation that was to influence the progress of biology until the Renaissance and beyond. Galen was the first experimental *physiologist* and a recognized authority on anatomy. He gave clear anatomical descriptions in his writings, and what errors he made were largely due to the fact that his dissections were restricted to work on pigs, dogs, and apes. Nonetheless, as a onetime physician to the gladiatorial school in Pergamum, he had many good opportunities for direct contact with the human anatomy. He was the first to show that arteries carry blood rather than air; he discovered and named many previously unknown muscles; and he did pioneer work on brain and spinal functions. After Galen's death in 200 A.D., little more was accomplished in biology for over a thousand years. His works remained the prime authority in anatomy until the time of Vesalius in the 16th century, and in physiology until the time of Harvey in the 17th century.

Biologists of the ancient world were usually either *physicians* or *philosophers*. The few practical, experimental achievements were generally made by the physicians, but new gains in knowledge were kept in perspective by the philosophers, who constantly critiqued the available data. Dissection was always a controversial matter, permitted at some times but not at others. Human dissections were frowned upon, particularly by the Egyptians, who believed that a corpse must be whole to gain life in the afterworld, and by the Christians, who believed in the sanctity of the human body, which was fashioned in the image of Christ.

There were very few medieval biologists, given virtual collapse of general education and academic pursuit during the long and dark Middle Ages. Religious institutions, the only remaining refuges of learning, were dominated by Christian theology, which acted against new knowledge in Western society. Many church leaders were convinced that science and religion stood in opposition to one another. And most of the academic activity that survived was strictly in the hands of the church—in monasteries, abbeys, and the like. Biology had a very small place in the curriculum that was developed at such places. Life was God's greatest mystery, the church taught, and people had no right to presume that they had the ability to understand it through either logic or experiment. If people were ill, it was God's way of punishing them for their sins; there was nothing more to it.

Biologists in other parts of the world did continue some interest in the science of life, although the greatest discoveries were destined to be made in the West at a later time. In the 11th century A.D., Avicenna presented the works and teachings of Hippocrates to the Arab world. Indian physicians showed a keen interest in the vital organs, especially the heart, which was thought to house the spirit. Breath was considered important because it supposedly forced sensory impressions through the heart, where they could be apprehended. Great emphasis was placed (and still is in modern India) on balancing a person's diet and behavior as a means of preventing illness, rebalancing being a cure for ailments. Today, proper nutrition and the alleviation of stress, as health factors, are only beginning to be given appropriate attention by Western physicians and biologists.

Chinese biologists made their own advances in the study of optics and in physiology. In the second and third centuries A.D. Chang Chung-ching, the "Chinese Hippocrates," believed that health was a matter of balancing life's vital fluids with respiration. Taoism inspired the teaching of breathing and self-control (concerning sex and diet) as techniques that produce good health. Tung Chi listed nearly 900 medicinal foods in the 11th century A.D., as the role of nutrition in health continued to be explored.

As Europe moved into its Renaissance, there was a renewed sense of scientific and academic curiosity. Renaissance biologists revered the works of the ancients,

Artists like these two were often important participants in botanical studies. (From De Historia Stirpium, *1542)*

Galen in particular, but did not make many original contributions at first. They were convinced that all that could be known about anatomy and physiology had already been learned. Nonetheless, the rekindled interest in the field was sufficient to bring biology to a point where it would one day question standard authorities and seek new horizons. At the turn of the 14th century Mondino de 'Luzzi finally dared to personally take up the method of dissection at the distinguished medical school at Bologna. Dissection had only recently become acceptable at the school; even then, it had be performed only in restricted cases and by lowly manual assistants—never by the *scholars* themselves. Mondino did his own dissections, though, and in 1316 he wrote the first book in history devoted purely to anatomy. His works and teachings earned him the title "Restorer of Anatomy." Regardless of the accolades, Mondino did not take issue with any of Galen's concepts, many of which were false. Some of the great Renaissance artists also dabbled in biology—particularly anatomy—as an outgrowth of their work with the human figure. Leonardo da Vinci, Albrecht Dürer, and Michelangelo all wrote treatises on anatomy and physiology.

In 1543, Nicolaus Copernicus made the first scientific break with Greek influence when he publicized his new findings indicating that the Sun, rather than the Earth, was the center of the known universe. This unleashed the Scientific or Copernican Revolution, which focused on reexamining the traditional ideas of classical science. In biology, the revolution began rather slowly, with most physicians and scholars clinging to orthodox views, while only a few dared to look beyond. Leonhard Fuchs's landmark publication on medicinal plants, *Historia Stirpium*, in 1542 redirected the field of botany. *Zoology* was given its modern impetus by the massive publications of Konrad von Gesner, the Swiss naturalist who wrote five volumes (heavily influenced by Artistotle) about different groups of animals. Ironically, Gesner's greatest and most original observations were made in botany, but his works dealing with them were not published until 200 years after his death.

The Belgian physician and onetime professor of anatomy, Andreas Vesalius, was among the first biologists to develop new techniques and to attempt to go beyond the works of their Greek forebears. Vesalius became a professor at the prestigious University of Padua when he was only 23 years old. There he became quickly disgusted with the crude dissections performed by unskilled assistants, and reinstated Mondino's earlier practice by performing his own dissections. He encouraged other biologists to do likewise. Vesalius proved to be a very entertaining as well as provocative instructor. But the substance of his lectures was often controversial and his findings unorthodox. In 1543 (the same year that Copernicus published his findings in astronomy) Vesalius published one of the most famous books in biology—and in all of science. Entitled *De Humani Corporis Fabrica* ("On the Structure of the Human Body"), it was the first really accurate anatomical survey. It diverged so greatly from Galen's views, however, that it was widely scorned in the profession. Vesalius lost his post at Padua, but later became the court physician to the Holy Roman Emperor, Charles V, and then to Philip II, king of Spain. He continued to be harassed by his opponents, who accused him of heresy and body snatching. To avoid execution, he consented to make a pilgrimage to the Holy Land to purge himself of sin, but he died when his ship was wrecked on the return journey.

The work of Vesalius was significant to biologists in several ways. First, he demonstrated that Galen was not beyond criticism. Second, he stressed the merits of hands-on investigation and experimentation, at a time when most biologists were content merely to compare theories and ideas. Third, he showed biologists a powerful device that could be used to make their findings available not only to other scientists, but also to the general academic audience. This device was the printed word, which had been available for some time, but not put to use for these purposes. And not only could Vesalius's book be reproduced in large quantities, but the book also made liberal use of printed illustrations that are still exemplary for their clarity and rare beauty.

Anatomy quickly advanced after the publication of *De Corporis Humani Fabrica*. Pierre Belon worked out some rather advanced comparative anatomies. Gabriel Fallopius accurately described the tube (now called the Fallopian tube) leading from the ovary to the uterus, a structure that plays a significant role in human generation. Bartolemeo Eustachio rediscovered the ear-throat canal noted in antiquity by Alcmaeon but now called the Eustachian tube. So much new biological knowledge was accumulating that by the end of the 16th century there were clear-cut specializations among biologists for the first time. The main, separate branches of study that emerged included *anatomy, embryology, physiology, zoology, botany,* and *pathology.*

With so much advancement in the science of life, biologists soon found themselves involved in one of the great scientific debates in history—the issue of *vitalism* as opposed to *mechanism.* Simply stated, the position of the vitalists was based on the ancient notion that there was a special vital substance that made living beings altogether different from nonliving matter. Just what this substance was became a matter of debate. Some vitalists thought fire, some water, and so forth. Christian

Physicians like Andreas Vesalius made important contributions to the scientific study of anatomy and physiology. (Woodcut by Jan Stephan von Calcar, from De Humani Corporis Fabrica, *by Vesalius, 1543)*

thinkers tended to think of it in more mysterious terms, like "the divine breath." The mechanists, meanwhile, thought that the new evidence being supplied by biologists and other scientists (particularly *chemists*) indicated that living beings were composed of essentially the same material as lifeless matter. Vitalists argued that human life was too sacred and mysterious to be reduced to rationality or quantification. The mechanists argued that, in order to best serve mankind (as opposed to upholding religious precepts), biologists ought to recognize that the structure and workings of the human form can and should be understood in order to derive cures for disease and principles of health and longevity. In short, biologists were challenged by the mechanistic point of view to use every means possible to make their discipline relevant to the improvement of life on earth, rather than preoccupy their thoughts with the design of the world beyond.

William Harvey, a 17th-century English physician, was the first modern physiologist to break away from traditional doctrines in his field, as Vesalius had done earlier in anatomy. The new ideas that issued from Harvey's work were brought forth by his unrelenting faith in experimentation as the most reliable scientific method of investigation. He believed that mere speculation was an outmoded tool of the biologist, and his own findings contributed greatly to supporting that view. Harvey was a well-to-do physician who was much more interested in biological research than in routine practice. His greatest contribution to the field was his pioneer work on the circulatory system. He demonstrated for the first time that blood does not ebb and flow like the ocean tide (an established view, carried over from the Greeks), but rather circulates in one direction from the heart through the arteries and then back again through the veins. His experiments were so exact that he even calculated quantitatively how much blood was actually pumped by the heart.

Harvey's career was a landmark for biologists, and not just because it dethroned Galen as the ultimate authority in the field. More importantly, it underscored the broad significance of experimentation and quantification as methods of research not only valid, but also the most reliable for future biologists. Moreover, it gave fuel to the mechanists, who took Harvey's work as final proof that the human body was in fact a smoothly working machine.

His work stimulated a greater interest in biological research than had ever before been known—research that remains in progress even today. Descartes, the 17th-century French philosopher, was so impressed with Harvey's findings that he determined to formulate a totally mechanistic physiology. It was, however, ill-conceived and sorely lacking in scientific documentation based on direct observation, which it so extolled.

Following Harvey's death, biologists were left to ponder over the mechanics of life. A mathematician named Borelli studied life in terms of levers and forces, dynamics and statics. He called this venture *muscular mechanics*. Franciscus Sylvius likened the human body to a chemical device, and discussed both digestion and disease in terms of chemical models and balances. Others followed suit as biologists fell squarely into the mechanistic camp, a situation that broadly applies to even contemporary biologists.

While biologists were becoming more interested in experimentation and direct observation, a new device was being developed that greatly aided just this sort of investigation. The first popular compound *microscope* was devised by Galileo in 1610. It soon became the biologist's best friend in the study of cells and tissues. One of the very first scientific societies ever formed was a group of 17th-century Italian biologists who had gathered together purely for the study of life under the microscope. The society, called the Academy of the Lynx, met in Rome at the home of its founder and president, Duke Federigo Cesi. Cesi did some pioneer work in anatomical botany; so did Marcello Malpighi, who also detected the capillary systems that connect human arteries and veins, and was the first to discover the corpuscles in the kidney through which nutrients are absorbed. But most of the great work in the age of microscopy was done by men outside of Italy.

Nehemiah Grew and Robert Hooke were leaders in microscopy in England. Hooke was the first to use the word *cellulae* in reference to the self-contained box-like structures that he detected in cross sections of cork placed under the lens. Anton van Leeuwenhoek and Jan Swammerdam led the field in the Netherlands. Although Leeuwenhoek was Dutch, he was a member of the Academy of the Lynx in Rome, while most of his research—including his discovery in 1683 of bacteria—was published in England. The work of the biologists was still considered relatively insignificant by most commercial publishing houses.

The 18th century saw biologists continue to collect encyclopedias of data based on microscopic observations, experimentation, and increasingly refined quantification models. Stephen Hales, a British pastor, founded *plant physiology*, even analyzing the rate at which water passes through a plant system. (He later was able to measure human blood pressures using similar methods.) *Biochemistry* had become an exciting field, as Rente de Reaumur demonstrated in his studies of digestion. He showed that when a hawk (which swallows its food whole) regurgitates its food, that food is already partially digested without any mastication. Furthermore, chemical analyses of the regurgitated meat showed the presence of digestive juices or chemicals, proving that digestion is a chemical process. Albrecht von Haller, a Swiss nobleman, then described the chemical activity of *enzymes*. The works of the great French chemist, Antoine Lavoisier, the British scientist, Joseph Priestley, and the Dutch physician, Jan Ingenhousz, eventually combined to show the chemical relations between plants and animals; how plants give off oxygen that is, in turn, used by animals; and the critical role of light in the chemical processes of both plants (directly, through photosynthesis) and animals (indirectly, by the eating of plants).

Throughout this period, the biologist's profession had itself been changing. Most ancient biologists were either philosophers or physicians. They did not limit their studies purely to biology, much less to a specific discipline such as anatomy or physiology. The same could be said of almost all biologists up to the time of the Industrial

Revolution. Most of the professionals were people of means and very frequently they were court appointees, nobles, and aristocrats. Many were also physicians, who often abandoned or neglected their practices for their experimental research. From the Middle Ages, when universities first developed in Europe, biologists were frequently appointed to academic posts. Since the major interest of the universities before the Englightenment was in philosophy and theology, many biology professors were obliged to remain faithful to orthodox Christian and classical ideas, even when these might reasonably have questioned them. Experimentation and quantification relating to the science of life were roundly scorned, since vitalism—the chief scientific philosophy for centuries—taught that the secrets of life (i.e., of *vital essence*) were known only to the Divine and revealed only to the devout. Any attempt to unravel such mysteries intellectually, vitalists believed, reduced life to a mechanism little different from rocks, oceans, or other inanimate entities. Some of the more serious and progressive biologists were fired from university and court posts, were socially and sometimes politically exiled, and found little opportunity for publishing their controversial findings.

Not all biologists were brash upstarts challenging the authority of church and state. Most were landed gentlemen and aristocrats who dabbled in biology as a genteel pastime and were typically determined to avoid any controversial or revolutionary doctrines. The majority of biologists were supporters of the status quo, even many of those who joined biological and scientific societies designed specifically to avoid the intellectual and experimental restrictions of the church-dominated universities. Biologists were so closely bound to the existing political and social structures, that during times of revolt their societies were sometimes purged—and some members even executed (notably in the French Revolution)—for their reactionary views and support.

The 18th-century Enlightenment created a new atmosphere for biologists to work in, emphasizing the ability of human beings to understand life through scientific investigation. The purpose of this understanding was to improve the human condition on Earth, rather than to postpone all human reward for the afterlife. Although the Enlightenment may have given a new intellectual stimulus to biological research, the Industrial Revolution in the next century gave it focus and actual application. As urban society developed, many problems—from public health to food preservation—demanded the attention of hard-working biologists, devoted to their science and determined to apply it to the betterment of humanity. Many biologists were needed to staff research teams in scientific societies and at universities (which by now had become more worldly in outlook, and therefore more accepting of the role of the biologist in academia). As a result the profession soon expanded to include people of all social, economic, and political backgrounds and persuasions. The democratization of society at large affected biologists, removing the occupation from the control of the nobles and political-social elites. In this new atmosphere of progress, opportunity, and challenge, the profession finally emerged as a significant source of vital research in the modern world.

With all the new information biologists were gathering, there was a great need for some sort of classification system to organize the material. Carl Linnaeus provided just that necessity with his binomial nomenclature, which featured the classification of organisms according to genus (e.g., *Homo* for primates) and species (e.g., *sapiens* for humans). But the species were still considered by Linnaeus and his contemporaries to be static and unchangeable. Only a few suggestions prefigured Darwin's startling concept of the gradual *evolution* of one species into another. The French naturalist Georges Buffon believed that species may change slightly over long periods of time, and he even regarded the ape as "degraded man." He also thought that earth might have been in existence for as long as 75,000 years, and that life might have existed for more than half that period. Shortly afterwards, the Scottish physician James Hutton demonstrated geologically that the earth had indeed existed for many thousands of years beyond the 6,000-year limit believed in by many Christians. Hutton went so far in his understanding of evolutionary change as to begin work on a theory of organic evolution by *natural selection*. His death ended the project, and it was half a century before Charles Darwin finally developed and popularized the theory with his publication of *The Origin of Species* in 1859. In the meantime, a French naturalist named Jean Baptiste Lamarck offered the first coherent—though to modern scientists deeply flawed—

On a botanical field trip to Lapland, Carl von Linné (Linnaeus) adopted local dress. (Authors' archives)

theory of organic evolution. He believed that many traits, even those acquired through genetic mutation or otherwise, could be passed on to future generations, according to a principle he called the "inheritance of acquired characteristics."

The story of the theory of evolution is central to the history of the biologist's profession. Darwin's theory of evolution was based on the "survival of the fittest" idea, by which those organisms best adapted to their environments were considered the most likely to survive. Most species, Darwin noted, reproduce many more offspring than are able to survive. Only those with the strongest characteristics survive, thus assuring the biological continuation of those characteristics through a process called *natural selection.*

Darwin's theories were formally presented to the Linnaean Society in 1858. His ideas were profoundly shocking to orthodox Christians, who felt that the theories were contrary to the Biblical version of the crea-

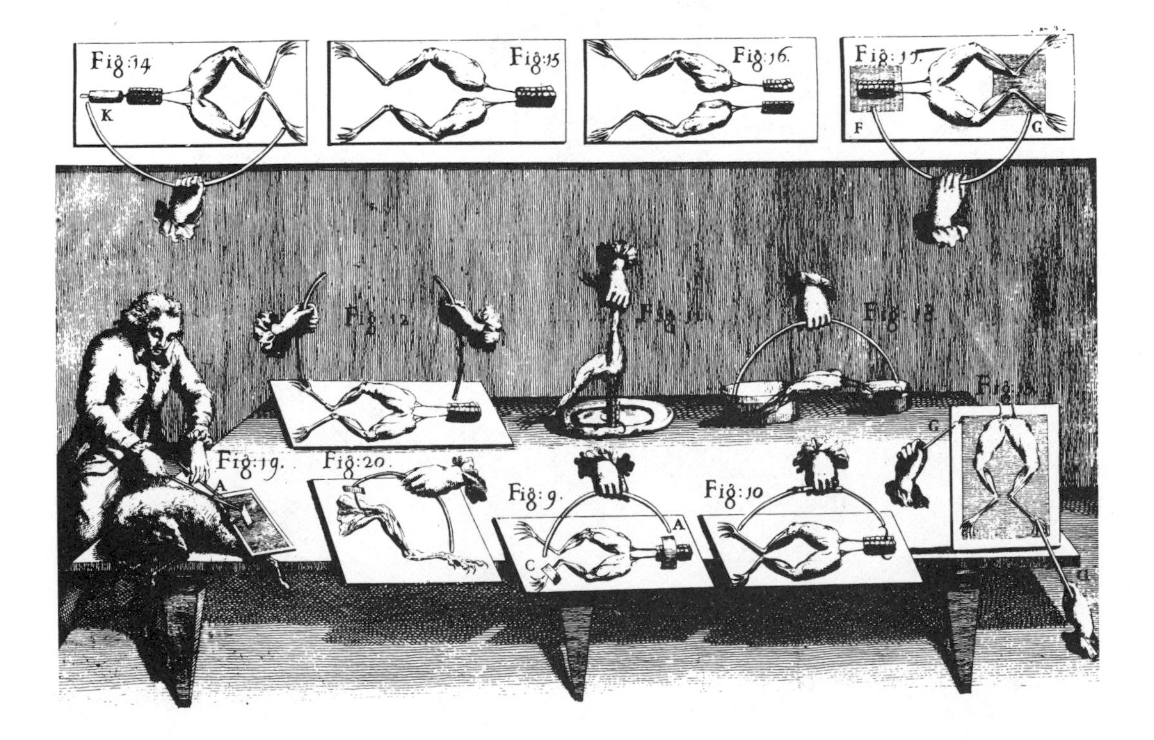

Biology and physics have long been closely related; this scientist is attempting to find electrical impulses in animals. (From De viribus electricitatis, *. . . by Luigi Galvani, 1791)*

tion of species. Others welcomed them as providing missing pieces to the puzzle of life—its origin and development. The theory of evolution was influential because it made many people believe that science was more reliable than religion in explaining the roots and destiny of human life. Today there is less of an intellectual separation between science and religion, but in the 19th century the perceived division stimulated further scientific exploration of life and matter. Darwinism even had social and economic repercussions, highlighted by Herbert Spencer's philosophy relating the survival of the fittest theory to society. Thus, it was thought that those in leadership positions in society belonged there because nature had selected them by virtue of their outstanding qualities.

Microbiology and *bacteriology* emerged as important fields of research in the 19th century. As early as the 17th century Francesco Redi had experimentally demonstrated that maggots that appeared on decaying meat had not been spontaneously generated from nonliving matter (the popular view for centuries), but had grown from eggs of microscopic size. It was not until well into the 19th century that the work of Lazzaro Spallanzani and Louis Pasteur finally disproved the theory of spontaneous generation. Pasteur later made an even greater contribution to biology when he detected microorganisms that he and his associate, Robert Koch, called *bacteria* (first identified by Leeuwenhoek two centuries earlier, but then largely ignored).

This discovery became especially exciting when it was learned that bacteria were the actual sources and causes of many diseases. In the 20th century René Dubos, a French-American *microbiologist*, expanded these findings with his elaborate experiments with *antibiotics*, drugs which act against the growth of certain harmful bacteria.

Biologists—increasing numbers of them women—have become more and more important as the 20th century has progressed, particularly in the fields of *genetics*,

In the wake of Darwin's work, biologists took considerable ridicule; this is a caricature of the Fellows of the Zoological Society. (By George Cruikshank, 1851)

biochemistry, cytology, endocrinology, ecology, histology, nutrition, and *molecular biology*. This has been an age of specialization and nowhere more so than in the field of biology. Some of the major specialties are discussed below.

The study of plant life was traditionally founded by Theophrastus of ancient Greece who wrote several hundred treatises on *botany*. In the first century A.D., the Greek physician Dioscorides augmented this work with his description of about 600 medicinal plants and herbs. Eastern *botanists* also made extensive use of herbal medicine and compiled extensive lists toward this end. In the same century an encyclopedia, *Historia Naturalis*, compiled by the Roman scholar Pliny, included some 16 volumes dealing with botany. The invention of the printing press in the 15th century led to an extensive body of literature about herbs, dealing almost exclusively with their medicinal uses. Most of these works were written by physicians who used such remedies in their practices.

The advent of microscopy in the 17th century opened new doors to botanists. They became more interested in describing the anatomy and physiology of plants from a scientific perspective; botany became largely divorced from the healing arts. The experimental investigations of

Stephen Hales put the profession on a new plateau after his intricate study of the movement of water in plants in the 18th century. Botanists soon became preoccupied with *morphological* studies (concerning size and structure), in addition to their *physiological* studies (concerning function). This trend culminated in the new, systematic classification of plants introduced by Carl Linnaeus.

Botany became extremely sophisticated in the 19th century, as scientists sought to discover evolutionary trends and genetic factors in plants. These investigations followed the work of Charles Darwin, who posed the theory of evolutionary change in species, and the Augustinian monk Gregor Mendel, who discovered genetic foundations to evolution in his studies of plant mutations. In the 20th century, botanists have been called on to help tackle the agricultural problems of supplying the world's exploding population with food. Botanists today apply their investigations to many other fields, such as biochemistry, physiology, molecular biology, genetics, and medicine. Most botanists are laboratory scientists working in universities and industries related to agriculture.

In contrast to the botanist, the *zoologist* focuses on animal life. Aristotle had made extensive observations and descriptions of many different animals and may be considered the first to show a serious interest in the field. Others continued this descriptive interest, but it was hardly a profession. Rather, it was an adjunct realm of interest among a few physicians and naturalists, or biologists whose main interest was in plants. Michael Scot and Albertus Magnus introduced the Aristotelian zoological classification to medieval Europe. But it was the systematic classification and binomial nomenclature devised by the Swedish botanist Carl Linnaeus in the mid-18th century that gave zoological research a solid and scientific framework within which it could expand. By the end of the century the French anatomist Baron Cuvier had established zoology as a real science. His intense interest in comparative anatomy led him to work in the classification of species, thus extending the work of Linnaeus. Cuvier compared present animal life with past life, but neither this nor his studies in comparative anatomy convinced him of an evolutionary doctrine. Instead, he championed the cause of *catastrophism*, the doctrine that only specific sudden events—Creation and the Great Flood, for example—caused changes in species. Nonetheless, his work along with that of Louis Agassiz, Lamarck, and others contributed to the mounting evidence in support of evolution.

By the 19th century, zoologists were counted among the professional staffs at universities and museums. Many are still employed in those capacities, but the profession is not as distinct today as it once was. Zoologists study animals in many special contexts. For example, zoologists who study animal genetics may be considered *geneticists* more than zoologists. This development has been caused by the vast specialization in the sciences that began in the late 19th century. The term *zoologist* today has little more meaning that the term *biologists*. People tend to speak instead of *physiologists* or *ecologists*, who study biological processes and relationships in animals but do so to gain a broader conceptualization of life in general—not solely of animals. This attitude has become particularly prevalent as the theory of evolution has become more accepted, allowing humans to be considered among the animal kingdom rather than distinct and separate from it.

Zoologists today very often specialize in studying a particular type of being. *Primatologists* study members of the primate species, often living on site for months or even years to observe communities of *Homo sapiens'* nearest relatives in the wild; *herpetologists* study the reptiles and amphibians; *ichthyologists* focus on fish; *ornithologists* specialize in birds; and so on.

Closely related to these specialties and shading off into the social sciences, are the *anthropologists*, who take as their charge the study of human beings. *Physical anthropologists* focus on the material evidence relating to early humans, especially in terms of their evolution into modern form and the beginnings of culture. Their study, in turn, shades into that of the *archaeologists*, who study the material evidence about the development of civilization. Traditionally archaeologists focused on studying peoples during the thousands of years before they developed written records, assuming thereby the province of the *historian*. But today archaeologists focus on physical remains of human activity up to and including modern times. *Industrial archaeologists*, for example, study the material record of the industrial era of the last two centuries. Another specialty that straddles several areas of study is that of the *cultural anthropologists*, or *ethnologists*; they observe and describe distinct ways of life, generally among peoples apart from the industrialized countries, paying attention to the interaction between people and their environment, as the *geographer* does from a different perspective.

Among the many biological specialties, *cytologists* study the internal organization and structure of the cell. By the 19th century it was accepted that the cell was the basic unit of all living organisms. Later in the century, cell parts were detected and described by several scientists. Cellular activity, particularly cell division, was noted soon after and formed the basis for the study of genetics. Max Schultze was the first to clearly describe the importance of protoplasm, the physical basis of all life. The investigation of cellular growth and division has been shared by many 20th-century researchers.

Geneticists study heredity, the passing of characteristics from parents to offspring. Units called *genes* contain the genetic information that ultimately determines which trait will appear. Geneticists study just how these genes are transferred from one generation to the next.

The 19th-century Austrian abbott Gregor Johann Mendel was the earliest pioneer in this field. He made elaborate genetic studies of peas at his garden in Brno, and was able to accurately predict which particular traits would appear in new crops and in what proportions. Hugo de Vries, a botanist from the Netherlands, then studied the nature of mutations, the unexpected breaks from an organism's genetic past. In the 20th century, geneticists learned how to artifically induce mutations with X-rays. The famous American geneticist Thomas Hunt Morgan learned that genes occur in linear order in the chromosomes.

During the 19th century geneticists were preoccupied with the relationship between mutation and evolution. Does the former explain the latter, or is Darwin's natural selection the chief cause of evolution? Most scientists in the field operate on the premise that both theories are credible and probably intertwined, and that neither excludes the other. Since World War II, scientist have paid especially close attention to the role of the substance called DNA (deoxyribonucleic acid) in the passing of genetic information. Geneticists today research *population genetics*, the effects of breeding and mutation on the genetic makeup of various living populations; *cytogenetics*, the study of how cell structure and the functions of chromosomes affect heredity; and *molecular genetics*, concerned primarily with the role of DNA in the genetic process. *Genetic engineers*, also called *agricultural engineers*, experiment with the isolation and development of genes in order to create gene cultures of desired traits that can be used to combat disease and breed superior varieties of plants and livestock. Some genetic engineers even postulate the possibility of creating an ideal human race through such techniques, although many moral as well as biochemical and genetic factors must be considered here.

Among the newer professions in biology is molecular biology, developed since World War II out of genetics and biochemistry. *Molecular biologists* are chiefly concerned with the interpretation of biological phenomena in terms of cellular molecules. They study bacteria, viruses, genes, enzyme (especially protein) synthesis, cellular regulatory processes, and the comparative molecular structures and functions of such proteins as hemoglobin and hormones. The findings of molecular biologists promise to be among the most exciting in the near future, offering important advances in fields such as medicine, nutrition, and agriculture. The profession is a new but vigorous one, similar in importance to that of the physicist in the first half of the 20th century.

Endocrinologists study the biological role of ductless (endocrine) glands and their internal secretions known as *hormones*. The mid-19th century French physiologist Claude Bernard was one of the pioneers in the field. In 1891 the British physician George Redmayne was the first to successfully treat myxedema, the common form of hypothyroidism (underactive thyroid), with an extract of the thyroid gland. Early in the 20th century, adrenalin, insulin, and cortisone were isolated by endocrinologists, whose findings were quickly used by the medical profession in combating a variety of disorders and previously lethal diseases, like diabetes. Their success has not been unalloyed. Patients treated may develop overdependencies on hormone injections or even suffer long-term side effects from them, as with the cortisone treatments commonly used to relieve rheumatoid arthritis. Also, the clinical use of hormones has been found to interfere with the body's natural tendency to achieve homeostasis, a natural chemical balance. Discoveries in endocrinology have certainly had wide impact, in any case. Endocrinologists have recently isolated and discovered sex hormones, used for birth control, among other purposes. And the marketing of birth control pills has had enormous social effect.

Histologists specialize in the study of living tissue. Tissues are groups of cells, similar in function and appearance, that make up plants, animals, and lesser organisms like bacteria. They were first so named by Marie François Bichat in the 18th century, but Marcello Malpighi, of the previous century, is considered the father of histology. Technological innovations, such as frozen tissue sections and electron microscopy, have permitted greater reliance on this profession by medical laboratories, hospitals, and zoological research centers or university departments. Histologists, like cytologists, work closely with *pathologists* in determining the presence of cancerous or diseased cell tissues.

Nutritionists study the processes by which plants and animals absorb and use food. They describe specific nutrients and identify their functions in the growth and maintenance of living organisms. Nutritionists advise *dietitians* on the proper use and combination of food in the diet, but more typically they are engaged in laboratory research. Their findings may be broadly applied to many other sciences, including orthomolecular medicine, which attempts to prevent illness through the use of natural substances—that is, food—rather than drugs or other organic stimulants such as hormones or glandulars (secretions from glands).

In a general sense, the history of the nutritionist goes back to the beginning of time. The Greeks, and especially their athletes, were particularly careful about the quality of their food intake. Cooks throughout the ages have studied the proper use of food and even their medicinal utility. But as a specialized science, the profession is a very new one. It may be said to have begun only in 1912 after Casimir Funk discovered vitamins, the coenzymes essential to proper metabolism.

Ecologists study the relationships between living beings and their environments. They usually study particular units, in which various living and nonliving things coexist simultaneously. The study of ecology can be traced as far back as Theophrastus and other ancient Greeks. However, it was not regarded as a serious field of inquiry until the 1960's when concern about environmental pollution and endangered animal species caught the attention of the public as well as scientists and politicians. Some people trace the roots of the modern profession to the early-19th-century work of Thomas Malthus, who correlated rising population trends with dwindling food

A soil conservationist and a staff biologist consult on water control for a wildlife refuge. (USDA-Soil Conservation Service)

resources. The Ecological Society of America was founded in 1915, but researchers have only recently specialized in the field. The recent concept of the ecosystem—a functional unit of various organisms interacting with the physical environment and conditions of a specific area—has given the field definite focus, and has permitted the development of the profession.

One more unusual biological specialty is that of the *taxidermist.* Taxidermy is an ancient profession typically associated with royal courts and the nobility. Taxidermists restructure and fill the carcasses of dead animals for display purposes. The art revolves around making the stuffed creatures resemble their living forms as closely as possible. During the Middle Ages, when hunting was a major sport and recreation in European courts, good taxidermists were widely sought and amply rewarded for their skills, which demanded a fair knowledge of anatomy and physiology. Although the occupation revolved around popular displays, it also contributed to an interest in the classification and comparative anatomy of animals, and thus to the development of zoology. Some taxidermists still make their living stuffing animals for sport *hunters* or *fishers*. But most taxidermists today are employed by museums and biological research departments in universities and other institutions. Taxidermists must be well-versed in anatomy and physiology as well as in the restructuring processes. In some cases they must do their work with only speculative information, as when they reconstruct a figure from a few remaining skeleton fragments.

For related occupations in this volume, *Scientists and Technologists*, see the following:
  Chemists
  Scientific Instrument Makers

For related occupations in other volumes of the series, see the following:
in *Harvesters*:
  Fishers
  Hunters
in *Healers* (forthcoming):
  Barbers
  Pharmacists
  Physicians and Surgeons
in *Scholars and Priests* (forthcoming):
  Scholars
  Teachers

Cartographers

Cartographers are concerned with the art of mapmaking. In early times, the Greeks were the first to make the art a science as well, using mathematical computations to derive the most accurate sea charts of the time. The *astronomer* Hipparchus was the first to use the famous grid pattern of latitude and longitude to better define areas. The Chinese, Indians, and, in the Islamic world, Moslems and Jews, went even further in this profession in terms of accuracy and detail. They used astronomical plottings in calculating directions and locations; their maps were used primarily by *sailors*, but also by overland travelers like *post-riders* and caravan spice *merchants*. Chinese cartographers produced the first printed maps, but their greatest innovation in the profession was the development of the magnetic compass toward the end of the Middle Ages.

Before the Age of Discovery in the 15th and 16th centuries, maps were generally of a regional nature, depicting areas such as the Mediterranean, China, or the Far East. As explorers took to the high seas in search of new lands and trade routes, *mapmakers* expanded their vision to include larger areas. The first true world map was not made until the late 19th century, but during the Age of Discovery the occupation flourished, as maps and sea charts guided many an adventurous voyager. It was during this era that Gerardus Mercator, the Flemish *geographer*, vastly improved the art of mapmaking with his innovation of *cylindrical projection*. This technique allowed the least possible distortion for *mariners* traveling long distances. (Distortion is unavoidable since the Earth is actually spherical rather than flat.) Mercator also drew up the first *atlases* (collections of maps),

In this 12th-century map, Islamic cartographers show their developing knowledge of the shape of the world. (By al-Idrisi, Bodleian Library, Oxford)

responding to the need for detailed regional maps that were related to one another in a systematic manner.

Cartographers had long been favorites of courts that spent a great deal of money sending armies and explorers to the far reaches of the unknown Earth. Many cartographers enjoyed royal patronage and favor. With the arrival of the industrial and scientific ages, the emphasis was more on accuracy than on postulating the possible locations of continents and seas, by then largely known. The profession became more of a science, and was greatly enhanced by a growing array of measuring and calculating instruments. The 19th century marked the great era of scientific expeditions, which provided new demands for cartographers.

Today, cartographers are aided by such advanced technology as computer-bearing satellites, which permit more objective views of the Earth's surface contours. The profession is extremely sophisticated, even in the development of areal mapmaking. Some cartographers specialize in mapping the galaxy beyond Earth, and even other galaxies beyond the Milky Way, the Earth's own galaxy.

For related occupations in this volume, *Scientists and Technologists*, see the following:

Astronomers
Geographers
Geologists
Mathematicians

For related occupations in other volumes of the series, see the following:

in *Communicators*:
Messengers and Couriers
Printers

in *Financiers and Traders*:
Merchants and Shopkeepers

in *Warriors and Adventurers*:
Sailors

Chemists

Alchemists dominated chemistry from ancient times until at least the 16th century, when the Copernican Revolution began. They actually continued working at the same time as the early *chemists*, into the 18th century and somewhat beyond. In many cases, the chemist and the alchemist were the same person, before the final split between alchemy as a pseudoscience and chemistry as an experimental and deductive science was fully realized.

The greatest contribution of ancient thinkers to the later work of chemists in the scientific mold was the search for basic elements in matter. The Greeks led in this search. Empedocles declared earth, fire, water, and air to be the four irreducible *roots*—what chemists would later call *elements*—of all that exists. Democritus, a contemporary of Empedocles, argued that all matter consisted of tiny particles that he called *atoms*, meaning "indivisible." All things represented varying combinations of these eternal particles, which were put into perpetual motion by the mechanical laws of nature. Although Democritus's *atomic theory* has been largely borne out by recent scientific experimentation, using modern technological devices and sophisticated quantification analyses, his view was unpopular in his own time and for centuries to follow. Socrates, Empedocles, and many others held the day, proposing all sorts of possible basic elements. Chinese philosophers, too, reasoned in terms of elements; a popular view in the first century B.C. listed five fundamental ones: earth, fire, water, metal, and wood. Perhaps the main reason that Democritus's view was ill-regarded was that it saw the world in mechanistic terms, making physical laws the reason behind matter. This directly contradicted the more typical worldview, which permeated ancient and medieval science, that the desires and plans of the gods motivated matter.

In the 13th century A.D., the English scholar Roger Bacon began to apply alchemical studies and techniques to medicine rather than to simply attempting to concoct gold (which he also experimented with). This interest in the alchemical aspects of medicine—a field known as *iatrochemistry*—was not fully developed until the time of the Swiss physician Paracelsus in the first half of the 16th century. Bacon had an even more lasting effect on the profession, however, in his rigid insistence on the scientific reliability of experimentation, observation, and mathematical modeling and quantification. As a result, he is often credited with having founded the *scientific method*. Before this, alchemy and chemistry had—like all pre-Copernican sciences—been distorted by theology, mysticism, and sheer speculation.

The study of chemistry was advanced in the 16th century by investigations into the mining industry, which was an important source of wealth for Europe's monarchs. Georgius Agricola's important work *De Re Metallica* was a milestone in *mineralogy* and the study of the Earth's chemicals. At the same time, Paracelsus was establishing an interest among chemists in medicine. Although this iatrochemical work relied heavily upon alchemy, the very notion that chemicals from the earth might be applied in a healing way to the human body opened chemistry to new perspectives. It also acted to support the mechanistic worldview that would very soon alter the whole scope of science. Two years after Paracelsus's death, the Copernican theory of the universe was published. With its emphasis on human reasoning abilities, rather than theological dogma, the Copernican ideas thrust chemistry into the midst of the broader Scientific Revolution that would begin to crystallize in the 17th century.

Many alchemists believed that mercury, sulfur, and salt were the root elements of matter. Jan Baptista van Helmont—a Flemish physician who, like Paracelsus, depended on alchemy in his practice of medicine—broke with this traditional view by proposing that water was actually the only basic element of the universe. With the Scientific Revolution developing, Helmont sought to demonstrate this belief rather than simply state it. He

Early chemists used distilling methods developed by alchemists long before. (From Liber de arte distillandi de simplicibus, *by H. Brunschwygk, 1500)*

did so by raising a willow tree which he nourished only with water; he showed that by the time the tree had gained 164 pounds, the controlled amount of soil had lost only two ounces. Accordingly, he declared that water was proven to have become a part of the tree's substance as it grew, thus being transformed into different matter altogether. Although Helmont's interpretation of the water-willow relationship was erroneous, it was significant that he sought to experimentally verify his water theory at all. In so doing he set chemistry farther along the road toward the experimental course, which would inevitably spell the undoing of alchemy. Although an avowed alchemist, Helmont is now thought to be the father of the science of *biochemistry*. Along with Franciscus Sylvius (some years later), he also began the study of the chemical basis of the body's digestion of food.

Although chemistry and alchemy had clearly become separate sciences before Robert Boyle was even born in 1627, the career of this illustrious Irish chemist is usually regarded as representing the final break between the two. Ironically, Boyle—considered by many historians as the father of scientific chemistry—was a well-known alchemist. But he was a renegade in that field. He did not believe that water, fire, air and earth—which many in the general population felt were the four basic elements of the universe—were elements at all. That is, he did not believe that they were unchangeable or indivisible. That distinction he reserved only for atoms, which he believed made up all elements. Furthermore, he believed steadfastly in the necessity of experimentation in the scientific investigation of the chemical and physical properties of matter. He preached that reason, rather than mystical speculation, would open up the unknown world of chemicals. By both model and example, he illuminated the way for his successors. He studied gases, guided by the views of Helmont, who had invented the term *gas*. Boyle attempted to detect smaller components of what had always been considered a single element, the air. He initiated the notions of *chemical analysis* and *chemical reaction*, and through these methods was able to propose some of the main distinguishing features among *elements, compounds*, and *mixtures*.

Boyle's work was profoundly important. He opened new territory for conquest and discovery by pioneering the chemical analysis of matter. In the century after his death, chemists followed his lead, formulating principles and discovering laws that would ultimately serve as the foundation of modern chemistry. Among those in this budding profession who were concerned somewhat with the actual science of chemistry were *physicians, geologists, mineralogists*, and *metallurgists*, who applied the principles of the science to their own disciplines. Many of these scientists were working *healers, mine operators*, or *prospectors*. Yet many others—and most illustrious of the lot—were aristocrats, or at least recipients of lavish patronage from aristocrats, and royalty. As chemistry became a more technical, practical, and honored science, chemists would gain more opportunities in diverse occupations in universities, governments, and private industry. Industry especially found use for chemists as the Industrial Revolution began to gain momentum in the latter decades of the 18th century.

The greatest controversy in chemistry during the 18th century—one that propelled the science to a new level of understanding and analytic potential—concerned the process of and the participating elements in chemical reactions. In particular, burning and rusting were examined by the greatest chemists of the period. They tried to discover exactly what chemicals and gases are involved when matter burns or metal rusts.

For at least a hundred years, it had been thought that a chemical or gas residing in the material being burned was the actual principle of combustion. One of the most eloquent proposals on behalf of this theory was made by the German chemist and royal physician Georg Ernst Stahl. He proposed that the more combustible a material, the more *phlogiston*—a term meaning "to set afire"—it contained. As a material burned, phlogiston was spent,

This late 18th-century chemical laboratory displays the ever-present beakers and retorts. (From Diderot's Encyclopedia, *late 18th century)*

and the same was the case in the rusting of metals. Although Stahl was obviously wrong, his daring analogy between combustion and rusting was quite perceptive and ingenious for the time. It gave future chemists an extra variable to use in the controlled experiments that eventually cleared up the dispute.

Stahl's phlogiston theory ruled the day until Lavoisier's dissent. Others sought to identify the principle of combustion, which they assumed lay within the chemical structure of the burning material itself. Joseph Priestley, one of the pioneers in the study of gases, developed a theory that explained the nature of the isolated gas which Lavoisier was to call oxygen. This gas, said Priestley, was best called "dephlogisticated air," because it had so little phlogiston that it rapidly absorbed that which was released by a burning material. The result was a faster and brighter-burning flame as the dephlogisticated air sucked up phlogiston. The basis of this theory actually recalled Daniel Rutherford's earlier studies of an isolated "phlogisticated air," which seemed so full of phlogiston that it would not accept that which was released from burning wood. Hence, the fire went out, in the presence of what Lavoisier soon after would describe more accurately as *nitrogen*. Priestley, meanwhile, had also isolated a gas which put out flames; it had already been termed *fixed air*, but soon would be known as *carbon dioxide*.

Oxygen, a key principle in a great many chemical reactions, had been isolated by many chemists. The Swedish researcher, C.W. Scheele, was probably the first who described *fire air*. But chemists also worked with many other gases. The English chemist Henry Cavendish called *hydrogen* "inflammable air" in 1766 and nearly two decades later was even able to produce water through the combustion of it. In the process, he demonstrated once and for all that water was certainly not an element, much less a *root element*. But for all the value of studies of gases like nitrogen and carbon dioxide, none of the studies could be fully appreciated as long as they were based—as all of them were—on faulty notions of the process of combustion, of the role that oxygen had in it, and of the role that gases in general had in the basic reactions of chemicals upon one another. Once these things were understood, chemistry could truly become a modern science.

The explanation of combustion, respiration, rusting, and the role of oxygen in all of them, was finally given by the eminent court scholar Antoine Lavoisier in the late 18th century. He applied one critical factor in his investigation of combustion and one distinctive mode of analysis, both of which had already been firmly grounded in the study of physics for nearly a century, but were slow to be taken up by chemists. The overlooked factor was weight differentials; the analytic tool was quantification. Lavoisier made startling discoveries regarding the nature of chemical reactions, because he was one of the first to realize the benefits of applying the new quantitative and mathematical calculations, which had so revolutionized physics and astronomy, to the otherwise relatively unused observations of even the sharpest chemists. As a result, his experiments had an exactness and indisputable character that no chemical theories had ever had before.

The principle of weight differentials—obviously dependent upon quantitative analysis if it was to be of any practical use in chemistry—was the foundation of Lavoisier's breakthrough in specific chemical analyses. It is one of the most significant general laws of both chemistry and physics to have been uncovered in the history of science. Specifically, it was his observation of weight changes in principal components of combustion that finally permitted his accurate interpretation of the phenomenon. His first clue came when he noticed that if he burned phosphorus or sulfur, the end products were heavier than the original substances. But if burning entailed the loss of phlogiston or any other substance, then the originals ought to have been heavier. Lavoisier then began to think that maybe the additional weight came from the addition of a gaseous component of the air, and that no substance like phlogiston was given off as a product of combustion. He tested this further by controlling the air supply and substance of a metal to be calcinated—heated—within a single closed container. While the container as a whole (holding both calcinated metal and air) showed no change in overall weight as the metal became calcinated, the relative weights of the metal and air were altered—the air becoming lighter and the calcinated metal heavier. Thus it became clear that some component of the air—oxygen, of course—had chemically reacted with and added its weight to the metal during the rusting process.

The summary of both experiments indicated that no matter how mass was redistributed, it could not be spontaneously created during a chemical reaction, nor could it be totally destroyed. It merely moved from one substance in the reaction to another. This, in effect, was the scientific evidence for the "law of the conservation of mass." This law inspired much 19th-century progress in the profession, as well as some of the great 20th-century discoveries. The law is still accepted today, although it was somewhat refined by Albert Einstein, who made great use of it in formulating his general *theory of relativity* in the field of physics.

Chemistry was not always a safe profession. Lavoisier, a symbol of the aristocratic powers who had controlled France for so many years, was executed during the French Revolution by those who sought to turn over the power of the state to a broader class of citizens. Moreover, by disassociating itself from alchemy, chemistry lost its mystical trappings and therefore whatever religious protection those may have afforded (little enough during the French Revolution). The chemists steadily adopted the new mathematics of Isaac Newton, the scientific methodology introduced by Bacon and Boyle, and most of all a *mechanistic* view of life and matter. The mechanistic view in particular brought criticism of the new chemists who, it was thought, had reduced life to a series of chemical links, events, and reactions that could be analyzed and manipulated through intellectual and experimental approaches. To many observers, this seemed to give humans a less than spiritual aura, exposing them to the laws of nature rather than to the design and guidance of the Creator. It was hard to accept, for instance, that the same process of oxidation that explained the burning of a piece of wood was also appropriate in explaining the physiological processes of respiration and digestion. It had always been assumed that a different set of laws applied to humankind than to the physical universe.

Were chemists not better than the *astronomers*, who had shown the Sun, rather than the Earth, to be the center of the solar system? Or the *physicists*, who had reduced the motion of the celestial bodies to such mechanical forces as gravitational pull? Or the *biologists*, who had studied anatomy and physiology as if they were the structure and parts of a machine, thinking little or nothing of such things as divine essence or soul? To the popular mind, evidently not.

Chemists continued their slide away from the religious

and the mystical realms in the 19th and 20th centuries. They had, after all, helped *geologists* to make extensive fossil studies that shattered the long-accepted notion that the Earth was only 6,000 years old. Eventually, this work led to Darwin's proposal that species were not created during a single event, as seemingly described in Christian and Jewish theology, but rather evolved over countless thousands of years to the point at which they now stand.

A good deal of the development in chemistry has been geared toward laboratory analysis and the accumulation of "pure" data. Very few elements were known in the early part of the 19th century. But since then chemists have isolated, identified, and analyzed at least 109 elements—and still counting. Laboratory chemists have also learned to synthesize and manufacture organic substances. Organic substances (those containing carbon) are found in living things, while inorganic substances (those containing no carbon) are found in non-living matter. Friedrich Wohler was the first to succeed in this venture; in 1828 he derived organic urea from inorganic ammonium cyanate.

The fact that Wohler and others after him were able to synthesize an organic substance from an inorganic substance eventually dispelled the doctrine of *vitalism* that many chemists had continued to hold well into the 19th century. Vitalism, a view championed by the followers of Georg Stahl in the 18th century, held that living substances were chemically different from physical matter—that is, that life was made of special *stuff* that differentiated it from matter. Sylvius, Paracelsus, and Helmont had effectively disputed this notion long before, as they worked to systematize iatrochemistry; they showed that, even more than a mechanical system (as the physicist Giovanni Borelli had argued), the human body was a chemical system.

But such views were not to be accepted until the experimental demonstrations of Wohler and other organic chemists. In the 19th century Justus von Liebig, in particular, synthesized scores of *organic compound and firmly established the study of chemical physiology.* Liebig also introduced the application of *organic chemistry* to new methods of *scientific agriculture.*

Others were busy studying the physical features of chemistry, such as the diffusion of gases, characteristics of colloids and suspension, and the ionization (electric charge) of solutions. The famous *phase rule* stated by the American chemist Josiah Willard Gibbs at the end of the century is studied by every schoolchild today; it refers to the three phases, or states, of matter: gas, liquid, and solid. In England Sir Humphry Davy was doing pioneer work on the effects of electrical currents on compounds. He was able to isolate a great many known and unknown elements using electricity. The French chemist Joseph Gay-Lussac showed that, when heated, different gases all expanded by equal amounts. This information was extended by the Italian physicist, Amedeo Avogadro, who went on to show that this meant all gases contained the same number of particles, given the same volumes. Avogadro's work was the basis of *molecular chemistry,* so named because the particles he studied became known as molecules. One of the greatest chemists of the day was Jöns Jakob Berzelius from Sweden. He published a textbook that was to have an unparalleled influence on the new field of inquiry. It was considered for many years the most authoritative work on chemistry available. Furthermore, he developed the use of chemical *symbols* and *formulas* to make laboratory work more practical and less cumbersome. Even laboratory techniques and instruction were vastly improved, largely thanks to Wohler.

Although cold, objective experimentation and quantification seemed to have turned the chemist into a *mathematician* more than a *philosopher,* there was still ample room for theories and hypotheses. Nowhere was this more evident or fruitful than in the atomic theories developed in the field of *inorganic chemistry.* The 18th-century scientist Torbern Olaf Bergman was one of the

In this laboratory in Giessen, biochemists Carl Justus von Liebig conducted his research on physiology and nutrition. (Deutsches Museum, Munich)

first to lay the foundation for modern atomic theory. He proposed the unique idea that there was a natural attraction of some substances to some others. This property he called chemical *affinity*. Although he labored over elaborate affinity tables, his ideas were soon forgotten; they were only recalled a century later, when they were equated with the interatomic force of attraction that seemed to bond elements and compounds.

John Dalton was the one who finally introduced a modern atomic theory that would help reshape the nature of chemical analysis. A poor lecturer who had all he could do to maintain order in his classroom, this early-

19th-century English schoolmaster was the first modern thinker to relate the notion of gaseous particles to a general atomic theory of all matter. In stating his "law of multiple proportions," he suggested that the same elements might combine with each other to form different compounds if they combined in different proportions. Thus, carbon and hydrogen on a 3:1 ratio formed methane, but the very same elements combined on a 6:1 ratio formed ethylene. This was the first time that a quantitative atomic theory became plausible, and in the following century it became the single most important theory in chemistry.

The identification and quantification of substances using an atomic model was made possible by differentiating the weights of each atom, as Dalton had clearly suggested. Theories of *valence* (the degree of combining power of a chemical element) were coming into being by the second half of the 19th century. They were based upon the relative atomic weights of elements. Berzelius had drawn up fine tables of atomic weights, and an Italian pioneer in the field, Stanislao Cannizzaro, used Avogadro's idea of molecules to note the importance of differentiating between atomic and molecular weights. Atomic weights were finally and clearly shown to be the key to a new era in chemistry.

But the idea was still only theoretically useful, for the actual arrangements of atoms in the composition of molecules could not be symbolized merely through the assignment of relative atomic weights. This problem was solved by German chemist Kekule von Stradonitz, who devised a system of structurally depicting the actual arrangement as well as the number of bonds to be found in a given molecule. Known as the *Kekule structures*, they were the last step in making atomic theory a workable method of understanding chemical reactions. Although this model was later elaborated and refined by Linus Pauling, it remains the basic formula used in deciphering chemical reactions today. By the end of the 19th century,

Dr. Harvey Wiley and his "poison squad" used to test food adulterants on themselves. (Washington, D.C., 1899, Library of Congress)

the chemists' general acceptance and use of atomic theory and valence finally enabled them to determine even the subtlest properties of chemical elements, and the most complex events involved in chemical bondings and reactions.

The life and career of the chemist had become rather an exciting one. Not only did universities offer prestigious chairs of chemistry, but there were also many exclusive scientific societies that welcomed chemists among their honorary members. Often a struggling laboratory or academic chemist might gain honorary membership in one of the top societies or academies by presenting some original theory or evidence to that specific scientific community, or sometimes by virtue of general accomplishment in the field over a period of years. As late as the 18th century a good many chemists still came from the aristocratic classes. This was natural, because only people of leisure could afford to spend significant amounts of time in the "idle pursuit" of scientific knowledge and expertise. Increasingly, however, chemists came to be recruited from the general population, as educational reforms gave more opportunity to those of middle- and lower-class origins.

Industrialization also increased the demand for chemists to work on innovations and refinements in private sector manufacturing. When the greatest surge of both industrialization and educational reform began in the 19th century, the two together provided an excellent climate for the burgeoning of the chemistry profession. It is not surprising that countries like Germany and France, leaders of the new educational reforms, and England and the United States, leaders of the Industrial Revolution, produced the greatest number of chemists. More important, among those numbers were the greatest pioneers in the profession. Outside of the Western World, where there was little progress toward either industrialization or educational reform, the chemistry profession barely existed.

The rewards of the profession remained largely the same for the greatest chemists. Many a renowned innovator of chemical theory or application was invited to take a seat in one of the great scientific societies, or sometimes to accept a lucrative position as a professor or librarian, which offered the opportunity to do research with little concern for earning a living. Beyond those offices, many of the French, German, English and

Scandinavian chemists were given royal patronage; some were even given titles by gracious and appreciating monarchs. Berzelius, at the age of 56 and considered the world's foremost authority in the field, was made a baron as a wedding gift from the Swedish king, Charles XIV. Sir Humphry Davy, raised in poverty, was granted British knighthood in 1812, after winning an award given by Napoleon for his production of an electric arc in 1805. Many observers thought that he should have not accepted the award since England and France were embroiled in war, but Davy insisted on separating science and politics.

Lavish honors and awards were heaped on the great chemists of the day and patrons offered professorships and court appointments. But the time had come when the profession was also to accommodate less spectacular workers. They were the ones who assisted the great scientists in their experimental works. Even more, they worked in private industries, which they helped in meeting the challenges of the new age of mass production and mechanization. That is not to say that the chemists who worked with private industries were necessarily of less stature than those associated with academia. Henry Bessemer invented a cheap and ready process for the production of a high-grade steel by the chemical conversion of phosphorus-free iron ore. The idea was put into action and Bessemer became an extremely rich man. In 1879 he was accepted as a Fellow of the Royal Society and was later knighted by the British Crown. Other industrial chemists made fortunes and earned high offices, too. But more important than this, in terms of the development of the profession as a whole, was the rising number of young chemists working with less praise and compensation. They worked in groups in order to pool their thoughts and produced some rather helpful, if not famous or revolutionary, concepts that played a significant role in spurring the industrialization of the West to newer and greater heights. Even Sir Humphry Davy, who made great contributions to theoretical chemistry and was well recognized and even knighted for them, was also a major innovator of mining technology. His *Davy lamp*, which made life much safer in the mines, allowed mining to make greater progress from that point on, and earned him the status of baronet for his service to the industry.

As chemistry progressed into the 20th century, its scientists earned greater and greater respect, improved their general economic status, and became more protected from the physical hazards of the job. (Many of the earlier chemists had seriously injured or even killed themselves by their contact with or mishandling of chemicals, or their unwitting breathing of toxic vapors. Davy, for example, had made himself an invalid in this way by the time he was 33.) Increasingly, chemists worked in teams, feverishly putting together new compounds and synthetic materials, or improving technology itself. The modern world began to revolve around chemicals; they were used to make clothing, highways, and even food. Concepts and theories continued to be worked on in the universities, but chemists were also greatly demanded by industry and the government, including the military, which sought more sophisticated and devastating ways to destroy its opponents.

New ideas also helped reshape the course of the profession in the 20th century. In 1895 Wilhelm Roentgen discovered *X-rays*, and two years later Joseph J. Thomson discovered *electrons*. Marie and Pierre Curie discovered radium in 1898; with Antoine Henri Becquerel, they pioneered in the study of *radioactivity*. Meanwhile, physicists were refining atomic theory even further to account for the actual structure and electrical charges of atoms. The place of electrons in chemical bonding was discovered during World War I by Gilbert N. Lewis of the United States. Soon after, the splitting of the atom became a reality. *Isotopes*, atoms of the same element having different masses, were discovered and studied. Progress in *synthetic organic chemistry* led to a burst in the manufacture of drugs, vitamins, hormones, and glandulars (substances secreted from glands). And thus, major chemical corporations came to have enormous influence over the medical establishment. Photosynthesis—plant production of nutrients using sunlight as the energy source—was first fully explained by Melvin Calvin, an American chemist. *Ribonucleic* and *deoxyribonucleic acids* (*RNA* and *DNA*, respectively) have been studied, particularly with respect to heredity. Extensive analyses of proteins and other *amino acids* may provide the most exciting and promising research in science in the years to come.

One of the prominent areas of scientific study in modern times has been biochemistry. *Biochemists* have been instrumental in many fields. These scientists are concerned with life as a system of chemical reactions and bondings. They study chemical processes that are fundamental to life. Modern scientists early noted chemical reactions within living organisms. The great French chemist Lavoisier, was among the first to consider the role of oxidation in digestion and the maintenance of body temperature. Biologists in the 19th century began to recognize and identify chemical enzymes, which act as

Managers and workers alike wear hard hats and protective goggles in dangerous working areas. (Cathy Holmes, from American Institute of Chemical Engineers)

specific organic catalysts in the maintaining basic body metabolism. The 19th century featured a classic dispute between the German chemist Justus von Liebig and Louis Pasteur. Liebig insisted that enzymes were chemicals, while Pasteur thought they were living organisms like bacteria. Liebig's point was finally substantiated in 1897, and during the 20th century biochemists have spent considerable time isolating and classifying specific enzymes and researching their unique and respective roles in living systems.

It was long believed that organic substances, which seemed peculiar to living organisms, could be converted into inorganic substances, but that the process could not be reversed. This assumption was based on the early-19th-century work of the Swedish chemist, Jöns Jakob Berzelius. After Berzelius's death, however, the German chemist Friedrich Wohler produced organic urea from inorganic ammonium cyanate, thus disproving the theory. Since then, and to this very day, biochemists have been quite preoccupied with the manufacture of synthetic organic compounds. The medicine, food, and pharmaceutical industries have supplied the main sources of employment in the field, besides university and government posts.

Closely associated with chemists in modern times are *laboratory technicians*. People in this profession are not always fully trained chemists, although they may be. Basically, they perform routine chemical analysis, sometimes on a massive scale and in a repetitious manner, but at other times in an experimental way that may require great skill, training, and precision. Laboratory technicians have been used by chemists throughout the history of experimental science, at least since the 17th century. At first, they were usually only the most prized students of master chemists who needed their assistance—and who also wished to act on a more personal level as mentor to one or two protégés. Many times, the student succeeded to the eminence of the master and became a full-scale chemist.

Chemical engineers use models to help them design plants for full-scale production of chemicals. (Diane Cook, from American Institute of Chemical Engineers)

In the 19th century, chemical analysis became extremely complex, as more data was gathered in the field and more appropriate chemical models were designed. At this point experimenters found they needed more assistance than they had ever needed before. They began to use teams of laboratory technicians for this purpose. Industry, especially, called on the services of large numbers of technicians to handle the large, diverse, and perpetual testing and experimenting that they directed. This continues to be true today. Medical and hospital laboratories also hire many technicians to perform routine analysis of blood and tissue specimens.

The work of the laboratory technician is extremely important, since the results of both chemical experimentation and medical diagnosis often depend on the accuracy of results handed down from the lab. At one time technicians were trained mostly on the job by the chief chemist or laboratory director. Large laboratories in the late 19th century began hiring chemists specifically to train technicians. Today, however, technicians usually receive their training at a special school, college, or university before they are employed, even though on-the-job training may supplement their formal schooling. Still other technicians are fully trained chemists.

For related occupations in this volume, *Scientists and Technologists*, see the following:
Alchemists
Astronomers
Biologists
Engineers
Geologists
Mathematicians
Physicists

For related occupations in other volumes of the series, see the following:
in *Healers* (forthcoming):
Pharmacists
Physicians and Surgeons
in *Manufacturers and Miners* (forthcoming):
Miners and Quarriers
in *Scholars and Priests* (forthcoming):
Librarians
Teachers
Scholars

Computer Scientists

Workers in the computer field design, modify, repair, maintain, install, analyze, operate, and store electronic computer systems and their input/output data. Computers are machines or devices that carry out complex mathematical operations related to specific data in such a way as to produce information readouts or computational results in intelligible and usable forms. Although the computer field includes some of the most promising and lucrative occupations in contemporary industrial society, it barely existed at all before World War II. Technology had not then advanced far enough to make computers available in any general or economically feasible manner. Certainly, there is a history of early attempts to make machines that would assist humans in calculating complex mathematical problems and in efficiently recording and printing out large masses of information. These early attempts set the foundations for the profession that finally began to open up on a large scale in the 1960's.

The first mechanical calculating machines and gadgets might be traced as far back as 3000 B.C. At that time, the Chinese toyed with a digital device they called the *Suanpan*, later translated into the Greek *abax* or *abacus*. The abacus was a primitive calculating board that allowed a user to store simple arithmetical information to be used in more complex calculating procedures. Since it was manually operated, however, it could not really be called a machine.

For a true machine, we must jump all the way to the 17th century A.D. It was then that Blaise Pascal made a shoebox-size machine that could carry out problems of addition and subtraction. It turned out to be quite helpful to Pascal's father, who used it for processing accounts for the family business. Several decades later Gottfried Wilhelm von Leibniz came up with a machine that also carried out the multiplication and division functions. The *Stepped Reckoner*, as it was called, was not made practical or useful until 1820.

The first direct, production-line application of such devices was made by Joseph Marie Jacquard. In 1780 he developed an automatic weaving loom that was operated from instructional punch cards fed into it. In the 1830's Charles Babbage attempted to make a mechanical digital computer that would have a *memory* (capacity for the storage of information) in the form of punched cards, along with a special unit (which he called the *mill*) in which to store data. But his *analytical engine* was far ahead of its time in theory, and the technology to make it successful did not then exist.

It would be a century later, in the 1930's, before Vannevar Bush would develop (at MT in Boston) the first automatic analyzer computer. His *differential analyzer* was capable of receiving and solving complex equations. The machine was all the more useful because of the mid-19th-century invention of *Boolean algebra*, which made it possible to formulate logical statements symbolically.

SCIENTIFIC AMERICAN

A WEEKLY JOURNAL OF PRACTICAL INFORMATION, ART, SCIENCE, MECHANICS, CHEMISTRY, AND MANUFACTURES.

NEW YORK, AUGUST 30, 1890.

THE NEW CENSUS OF THE UNITED STATES—THE ELECTRICAL ENUMERATING MECHANISM.

Computing devices aided in analysis of the U.S. census as early as 1890. (Scientific American, *August 30, 1890)*

This allowed people to deal not only with mathematical problems, but also with verbal and logical concepts in a systematic way.

After Bush's work, computer technology advanced

rapidly. The first semielectronic *digital computer* arrived in 1939, and the first fully electronic one seven years later—the ENIAC, developed at the University of Pennsylvania. In 1944 the Mark I had become the first computer system to employ electromechanical *relays* (switching devices), but the completely electronic ENIAC in 1946 proved to be a thousand times faster in its processing and computing. The first digital computer to have the capacity to store coded programs in a *memory system* (first envisioned by Babbage) was developed by John Von Neumann in 1949. Two years later a modified version of the ENIAC became the first widely marketed computer. With its ability to process both alphabetical and numerical codes, it opened a few frontier for the computer science field, and started a revolution in information, calculation, and processing systems that would eventually affect virtually all institutions.

Up until this time, there were very few professional people in the field. The *physicists, mathematicians* and university *professors* who spent some of their time working on such devices could perhaps be considered professional *computer scientists* in some sense. But it was not until the ENIAC and its companions began to be marketed in the 1950's that there developed a real need for design, production, analysis, service, and operations personnel. *Second generation computers*—those made faster, cheaper, more compact, and more reliable by the invention and refinement of transistors and solid-state devices—soon hit the market. These new machines created an even greater need for computer specialists in this field with its growing list of new computer users.

Still another surge in activity within the industry came with the sudden development in the 1970's of *third generation computers*. These were the miniature and cheaper-than-ever machines made possible by the development of the integrated circuit and the microprocessor. The third generation computers made computer systems available not only to small businesses and agencies, which could not have afforded them before, but also to the general public, who could now purchase cheap pocket calculators and television computer games. Schools began purchasing computerized educational equipment for use in the classroom. Everywhere one looked there was a market for computers at some level.

Today the demand for computer *designers*, *analysts*, and *operators* is greater than ever. Although computers have replaced human labor on many levels, they have also created plentiful jobs for computer specialists who have been trained in the field. There are many classifications of workers in this field, covering a broad spectrum of employment situations. Computer specialists work for the police, the army, national tax collection agencies, census takers, pollsters, space programs, nuclear energy and other utility corporations, and weather bureaus. They work in all kinds of businesses from major manufacturing, banking, and retailing firms, to the small retail stores and warehouse operations.

For convenience, computer specialists may be thought of in terms of three broad categories: *computer engineers*, *computer programmers/analysts*, and *computer operators*. The engineers construct and design improved models as well as initiate new applications for computers. Programmers and analysts are involved in setting up systems; finding the most efficient use for particular programs or systems, given a specific problem or application; and teaching users how to make the most of the system. The operators oversee the physical processing of data and run the machines during normal work days. They have little to do with what is put into the system or what comes out of it, except in a clerical capacity. Engineers, programmers, and analysts must have extensive training and education, especially at the engineering level. Operators, however, often have no special training except what they receive on the job. They are modestly compensated for their labors, while engineers, programmers, and analysts often find the field quite lucrative.

Programmers and analysts also enjoy considerable recognition and status, owing mostly to the mystique of their special knowledge. They use obscure computer languages and lingo that are little understood by the actual users and operators of the systems. In their work, they introduce startling changes in the daily operation of businesses and agencies. Their efforts completely overhaul manual systems of payroll, accounting, even production; in the process, they drastically reshape the use of labor. Engineers, too, enjoy considerable status, although their jobs are less visible than those of programmers and analysts.

More specifically, *computer engineers* plan new systems and modify the old ones, based upon their expert analysis of particular applications required for specific projects and workloads. They must determine the most appropriate data-processing systems required in given situations. They must then adapt existing systems to those situations—or sometimes devise wholly new systems that are tailor-made to fit unique needs.

In making specific applications of computer systems, engineers must formulate mathematical and symbolic models of the systems, designed to solve targeted problems or supply required information. In the process, engineers work closely with those who will actually use the system, in order to properly identify their needs and expectations. Based on such consultations, they then draw *data-flow* charts mathematically devised to meet the needs of the client.

Computer analysts set up computer systems in such a manner that they may logically and mathematically solve technical problems. They also devise data-flow charts designed to make data-processing systems responsive to specific needs. Simply stated, their job is to reduce complex problems, such as the control of a manufacturing process, into computer-processible forms. They are very concerned with the cost-effectiveness of the systems they recommend and install, and work to bring cost in line with the budget of their clients, without sacrificing overall quality.

Computer programmers work with the computer systems that engineers and analysts have determined to be best suited for a given problem or work load. Programmers may devise business/industrial or scientific/engineering applications for the systems. The programs they develop entail elaborate systems of instructions and data-processing routes, designed to deal with problems and work functions in as simple, efficient, and usable a manner as possible. The programs themselves are written in highly technical and specialized languages, such as FORTRAN or COBOL.

Programmers must give the originator , or client, complete instructions on how to code and enter information to the system, and how to decode the results. They work closely with the originator to be sure that every detail of the program is complete and manageable, so that the installation may fulfill its anticipated role in carrying out a specific application. The nature of such applications varies widely, from the commonplace, such as processing payrolls or college registrations, to the unique, such as building a bridge or spaceship equipment, or developing new scientific equipment for use in forensics.

Since the programmer works so closely with the originator of the problem or demand, he or she is the most visibly responsible for the contribution that a computer may make in a given situation and to a specific application. Moreover,the language used to achieve such results is usually poorly understood by the originator, who must rely heavily on the programmer for program maintenance and revision. This situation not only lends programmers a distinct air of authority and expertise, but also allows them to demand high fees for their work.

In case of small businesses, a single programmer often controls a computer system which may have become the backbone of operations. If anything goes wrong, he or she must be called upon—and paid whatever is asked for. Some disreputable programmers go so far as to use their own individual "languages" in programs to prevent their decoding by anyone else—even another programmer. In such cases they are free to disturb a system to the point

Professors of computer science are often highly sought after by industry. (The Institute of Electrical and Electronic Engineers, Inc.)

that it must shut down. They then pick up the necessary repair fee. Such cases are isolated, of course, and in general programmers enjoy high social esteem as well as financial rewards. These rewards have increased with the spread of small computers using prepackaged programs, or *software*. Programmers today sometimes work on their own or for software companies, writing programs for distribution to thousands of buyers.

Computer operators are essentially clerical workers who handle the computer's controls. They handle the input and output of data pertinent to business, science and government needs, so that it may be readily available for storage and immediate use. They learn to operate a wide range of instruments and control panels and must accurately record operating and down times. Some *terminal operators* work on-line computer typewriter or keyboard terminals, which transmit to or receive data from remote locations. They must know the coding system, since they are required to type input data. They also place requests for output data, which is received in the form of a printout or a viewing screen readout. Small-scale operators work desk-size computer equipment.

Some computer operators specialize in running peripheral equipment, which complements the main console. They are often given the title of their specific operation, such as high-speed printer operator. The equipment they deal with may involve external memory, data communicating, synchronizing, and input/output recording or display devices. These jobs usually deal with the physical aspects of handling computers, such as mounting and positioning reels on spindles, or preparing equipment for use and checking malfunction indicators. Computer operators have about the same financial standing and professional status as other clerical workers. Their training is usually obtained on the job and requires little education.

Other specialists also work in the computer field. *Computer technicians* help design, test, troubleshoot, repair, and modify computer equipment. Typically, they are called in to repair malfunctioning systems. In some cases, as in hospital laboratory analysis systems or national security projects, computer technicians are called in to make emergency repairs that are of an urgent nature. In such cases, they often work day and night in teams until the problem is solved and the system is once again working properly.

Tape/disk computer librarians classify, catalog, and maintain computer tapes and discs in special libraries designed solely for that purpose. Their main job is to make computer-processed data readily available for use at any time that it is needed. They are specially trained in classification systems.

Computer sales specialists are like any other salespeople in that their chief aim is to convince clients to buy their product and service. What distinguishes them to some extent is the considerable amount of technical

Systems analysts review needs and goals and design computer systems to meet them. (Association for Systems Management)

knowledge they must possess in their field. They are usually required to have an advanced understanding of engineering, electronics, and mathematics, as well as some notion of the business or scientific needs of their clients. If successful, they may earn very substantial incomes, owing to the high prices of the computer systems they sell.

For related occupations in this volume, *Scientists and Technologists*, see the following:
Engineers
Mathematicians
Physicists
Statisticians

For related occupations in other volumes of the series, see the following:
in *Financiers and Traders*
Merchants and Shopkeepers
in *Scholars and Priests* (forthcoming):
Librarians
Scholars

Engineers

Engineers generally put into practical use the abstract knowledge of the scientist. Few professions have portrayed *homo faber* (man the maker) in such clear and dramatic form. In early agricultural villages engineers may have been employed in the building of crude roadways and irrigation ditches. However, the first truly professional development in this field came in the person of the *military engineer.* As civilizations became more stable and began to occupy fertile valleys and accumulate wealth, the greed for riches and territorial rights emerged as a prime motivator of human behavior. War became the chief means of dispossessing and plundering one another.

In an age before explosive weapons or firearms, hand-to-hand combat was the main means of waging war. Village dwellers, rightfully concerned about attacks against their properties and persons, retreated within fortified citadels—cities with walls to keep out invading *soldiers*. The walls ensured sufficient safety until mass-destruction weapons (like gunpowder and bombs) were invented. Some successful invasion techniques against the citadel were to surround it with armies who would try to scale the walls, or smash through the massive city gates, or wait until provisions ran out inside and those besieged had to fight, surrender, or starve. This so-called siege warfare prompted the rise of the *military engineer*, who virtually dominated the profession until the modern Industrial Revolution.

War was a way of life in the ancient world, and the military engineers played a central role in it. But their knowledge and technique also made them *civil* and *mechanical engineers* dealing with public works and machines, respectively. They had to know basic principles of mechanics in order to plan and build devices of war, such as catapults and ballistae used to hurl missiles, (generally arrows or rocks). They often had a rudimentary understanding of hydraulics, since they were obliged to drain marshes for seaward defenses and for road or fortress construction. But above all, they had to understand the principles of construction in order to erect forts, build earthworks, and develop road systems. Despite the need for all this knowledge, most engineers developed their skills through trial and error more than formal education or apprenticeship. Although many had some working knowledge of arithmetic, geometry, and physical science, most relied purely on common sense and persistence.

One of the earliest engineers was Imhotep of Egypt, the chief director of the construction of the stepped pyramid at Saqqarah, built around 2250 B.C. Egyptian engineers were usually *master builders* in the service of the pharaoh. They enjoyed considerable prestige for their expert work in the construction of great pyramids and palaces as well as extensive irrigation and canal systems, which allowed the Nile to fertilize the surrounding

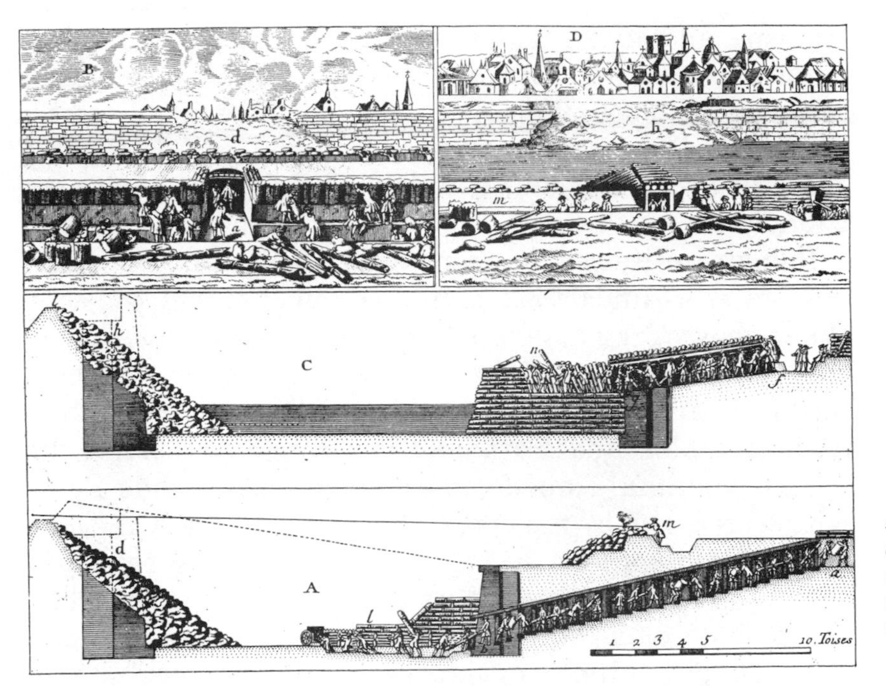

Military engineers designed and supervised the building of the fortifications that dominated warfare for many centuries. (From Diderot's Encyclopedia, *late 18th century)*

desert. Mycenaean engineers built great earthworks to control floods, and Mesopotamian courts also retained master builders for large public works. But the most important and extensive work of the ancient engineer was done in service to military and defense demands. In the seventh century B.C. the Assyrians were constructing moving towers and wheeled battering rams; and in the third century B.C. the Chinese began one of the greatest engineering tasks in history—the building of the Great Wall of China, which eventually stretched some 1,500 miles to protect the empire from foreign invasions.

Military engineers undertook many tasks that could be considered civil works, but were usually undertaken for military reasons. The Greeks and especially the Romans built advanced road systems, not to promote commerce or travel, but for the movement of troops and the transport of provisions. The peoples of the Mediterranean, notably the Greeks and Phoenicians, were the first to build harbors, in conjunction with the development of naval power. The Romans were famous for their excellent engineers. Crassus, Trajan, Constantine, and others undertook the systematic training of engineers, who built roads, bridges, lighthouses, and seaward defenses. Roman camps had to be protected by ditches and ramparts, and the Romans built Hadrian's Wall to protect over 70 miles of Britain's northern boundary.

Roman military engineers were inevitably the same officers who undertook civil and mechanical projects. When they were not accompanying military expeditions, they were put to work by the Republic or the emperor building amphitheaters, like the Colosseum, public baths, temples, and drainage and sewer systems, like the Cloaca Maxima. Aqueducts were built to provide Rome with water, and by the end of the first century A.D. some 300 million gallons were flowing into the great city thanks to the expertise of its engineers. As with public buildings, roads, and other such works and edifices, maintenance of these aqueducts was the responsibility of the Senate. The *statesman* Agrippa once ran a training school for *slaves*, who might thereby learn the art of designing and maintaining aqueducts. Despite the training of slaves for the profession, their duties must have been generally restricted to the aiding of the masters. Vitruvius, the most celebrated of all Roman *engineer-architects*, in his famous *De Architectura* (written at the end of the first century B.C.) noted that those in his profession had to be well born and trained meticulously. He also attests to the great importance of the engineer as a military officer and master constructor. The high esteem in which these roles were held is unsurprising, for the Romans held military prowess to be central to their notion of civilization.

Roman engineers greatly refined the business of military and civil construction. They followed the lead of the Greeks in building harbors and making aqueducts through mountains ranges to feed into sophisticated water supply systems. But the Romans, pressed as they were by the urgency of war, and inclined as they were to practicality in all matters, developed new techniques, experimented with a wider variety of materials, and introduced a greater economy into engineering than had previously been known. Supported by government funds and slave labor, Roman engineers efficiently and economically went to work in constructing the best roads, water systems, harbors, and earth defenses that would be seen in the West for many centuries to come.

Many of the Greek and Roman inventions and construction processes were forgotten or put to ill-use in the Middle Ages. The gear, screw, rotary mill, and water mill were known but unimproved for many centuries in a world that fell back into a strictly agrarian way of life. The greatest strides in the profession during these Dark Ages were taken in the East. Perhaps the greatest single contribution was the Syrian development of the mathematical concept of *zero* (inspired by earlier Chinese and Indian studies), which permitted a far greater refinement in construction planning and design. The Moslems did the most advanced mathematical work of the time, while Chinese and Indian engineers made the greatest application of these principles in both civil and military works. China boasted elaborate cities and advanced water systems during this time, while the Hindu Renaissance in India featured such wonders as the temples at Elephanta and Ellora.

The concept of *perpetual motion*—of a machine that would continue working indefinitely without continual addition of energy—was first developed in the middle of the 12th century A.D. by the great Indian scholar, Bhaskara. From that time on, many engineers were entranced with the idea of building a perpetual motion machine. Though all such attempts failed, their efforts greatly contributed to the development of military machines and missiles. Mechanics was not greatly enhanced beyond this point by Eastern engineers, although it is thought that the *windmill* was originally inspired by the Tibetan prayer wheel, and the waterwheel was greatly refined and reintroduced into Europe through Spain by the Moslems.

Eastern engineers almost single-handedly kept the engineering profession alive during the long Middle Ages, when technical advances stagnated in Europe. Although closely related to military operations, a great many of the Eastern engineers were civil employees, and as such were usually well-trained in religion and the arts as well as in purely mechanical procedures. Chinese, Indian, Japanese, and Moslem engineers were supposed to contribute not only to the technological progress of society, but to the spiritual and artistic side of life as well. Their work was not purely functional, even in times of war. Rather, they sought to attain for it a balance with, and thereby draw energy from, the natural environment of which their work was a part. If an engineer's work met these high ideals and ethical standards, he was among the most revered people of his time. For the most part, however, since many Eastern peoples were unimpressed with purely practical and technical achievements, the average engineer was regarded less reverently, as merely a functional civil servant or as a necessary but inglorious member of the military.

In Europe, the profession continued to center on the art and technique of siege warfare, eventually to be rendered obsolete by the Chinese development of gunpowder and firearms. In the 12th century A.D., a famous British engineer of his day—Ailnoth—worked in the Tower of London as a *carpenter*, as most men in the profession did then. The title *engineer* was apparently little used before the 19th century. Ailnoth, like his contemporaries, was referred to as an *ingeniator*, one who tried to invent new materials, machines, and techniques of war. Gerard, another famous British military engineer of the period, who served Henry III in the 13th century, was referred to as a *technical officer*. A sketchbook of Gothic engineering was written during the same period by the Frenchman Villard de Honnecourt, who noted the great amount of skill and training that characterized the profession.

The Western engineers of the 12th and 13th centuries

A.D. were preoccupied with the construction of defensive structures, such as castles, drawbridges, towers, and moats, and with the development of the great war machines (catapults, ballistae, and crossbows) that were widely used before the appearance of gunpowder. Shortly thereafter, when gunpowder came into general use, the military engineer became more involved with war gadgets and devices than with great earth and edifice works.

The Renaissance engineer was a person of varied talents—an *artist*, an *architect*, and a *scientist* as well as an engineer—who might be engaged to do intricate mosaics in a cathedral, sculpt a bust, construct a bridge, devise a sewage or drainage system, or besiege a town. Leonardo da Vinci was the prototype par excellence of the Renaissance engineer. He was one of the first engineers to work with an extensive knowledge of both mechanics and hydraulics. He had a keen insight into the nature of the materials with which he worked, and a strong background in mathematics, especially geometry. He devised many gadgets and machines, and even proposed a flying machine, based on a diligent review of the aviation techniques and physiological systems of birds. Most of his wonderfully conceived inventions did not work because of the limitations of mathematical and scientific knowledge at the time, or because of the lack of appropriate materials and construction techniques. Nonetheless, the vast scope of his interests shows how widely versed engineers of the period were.

It is not presumptuous to consider Leonardo an engineer as well as an artist; while some *painters, sculptors*, and *architects* of the time received royal or noble patronage for the continuation of their efforts, most found paid employment first and foremost as military engineers, since this was the most practical and marketable aspect of their talents. In attempting to persuade the Duke of Milan to patronize his work, Leonardo himself once made a lengthy plea, describing his achievements and expertise in constructing military devices, earthworks for defense, and roads and canals for the movement of troops. It was only at the end of the plea, and almost as an aside, that he humbly added: "In painting I can do as well as anyone else."

Kings and nobles would readily house, support, and reward engineers for their contributions to winning wars, but were less likely to support the fine arts or civil engineering. Although the growth of towns and cities at the end of the feudal age increased the need for roads, sewers, public buildings, and city planning, there were still very few opportunities for engineers to work within the civil realm. Most engineers, then, still sought patronage through paramilitary service.

The training of engineers became more scientifically based after the 17th century with the advent of Galileo's law of inertia, Newton's laws of motion and mathematics, and the generally heightened sophistication and application of both science and mathematics. But nothing had the effect on the profession that the Commercial and Industrial Revolutions did. With siege warfare virtually replaced by open-field combat between troops by the 17th century, there was a diminishing, less intensive need for military engineers. Many of these turned to the making of weapons and firearms. But most cast their sights on the new trade, business, and production activities that were turning Western Europe into the economic leader of the world. This is not to say that military engineers were no longer important. Sébastien Le Prestre de Vauban of France founded a prestigious school of military engineers in the second half of the 17th century, and the traditional French leadership in the field endured for quite some time thereafter. When George Washington established the Army Corps of Engineers in America in 1779, he enlisted the aid of the French. Military engineering is still important today.

Generally speaking, activities within the engineering profession were gradually turned away from the battlefield and into the civil world. The first school of civil engineering, marking the beginning of the modern engineering profession, was founded in France in 1747 as the *Ecole des Ponts et Chausees* (School for the Construction of Bridges and Highways). It gained widespread acclaim for its program of instruction in mechanics and design innovation. In Britain, at the same time, only military engineers were given the professional title of *engineer*, until well into the 18th century, when John Smeaton, designer and builder of the Eddystone Lighthouse, first called himself a *civil engineer* to distinguish himself from the military builders. The British Institute of Civil Engineers was finally established in 1818, followed by the Institute of Mechanical Engineers nearly 30 years later.

The Commercial Revolution had provided the first great impetus to civil engineering. The growth of towns and trade in Western Europe gave rise to the *millwright*, who designed and built grist mills to supply townfolk with cut grain, flours, and meals. *Roadbuilders* found better and more efficient ways to improve overland travel, so that *traders* could have better access to the markets of the towns springing up in Italy, Germany, Holland, France, and England. Municipalities themselves, with their public buildings, churches, and roadways provided civilian employment opportunity for many engineers. Still, civil engineering was not yet a specialized trade. Carpenters, artists, and various craftsmen all tried their hand at it, along with the always-present moonlighting military engineers.

The great period of professional growth in the field came between 1750 and 1850, the early years of the Industrial Revolution. It was during this time that machines and engines came to replace manual labor, and large factory buildings had to be constructed to house this heavy equipment and sponsor its related production activities. In earlier times, whatever mass manufacturing had been done—notably in the woolen industry—had been carried out in private homes, usually in the countryside. Although this so-called domestic system was quite widespread in England before the advent of the factory, it was in England that the Industrial Revolution first burst through the old manual labor and small-scale production economy. Not surprisingly, it was there that engineering enjoyed its greatest growth and some of its most memorable achievements.

The demands on industrial mass production to serve an ever-growing urban society led directly to the great engineering innovations of the age. The need for more yarn and cloth, for example, led to the development of advanced textile machinery. The demand for more coal to run the factories inspired the invention of the steam engines. They were used first for pumping groundwater out of the mines themselves, but were quickly adapted to direct industrial uses such as blowing air into furnaces and hammering iron. They eventually came to replace the waterwheel as the main source of industrial power or energy. Increased production activity meant that cheaper and better forms of transport had to be developed for reaching eager markets. While some engineers designed factories and mills, and others designed, invented, and built machinery and engines to make them work, still others busied themselves with the large-scale construction of ports, canals, roads, and bridges to carry the resulting great volume of manufactured goods to their destinations. The earliest mechanical engineers of the industrial age were, necessarily, inventors of new gadgets, machines, and engines. They had to devise new machines and techniques that could effectively and economically substitute for human sweat, painstaking labor, and piecemeal production.

Also important were the *mining engineers*, who for centuries had been making it possible to probe deep underground to obtain rich resources of mineral and rock wealth. Although mining engineers had been important as far back as the Middle Ages, they became more so during the industrial age, when coal became the leading source of energy for the great factories, first of Britain, then of Germany, France, and America. Underground mining was an intricate and treacherous business in Continental Europe, where the earth was little tolerant of deep or wide shafts. The situation was complicated all the more by the vast system of existing ancient works, which had to be shored up or avoided by engineers. German mining engineers rightfully earned reputations as the finest in the world. Although Britain's terrain posed much less a problem to miners, many a British king paid large salaries to teams of imported German mine engineers.

The 19th century saw an expansion of the social applications of engineering. The provision of public water supply, sewers, and city sanitation became the responsibility of *municipal engineers. City planning* became a major occupation in the industrialized states, and by the middle of the century transportation had become one of the leading areas of concern in the profession. The development of railroads and extensive highway systems, demanded the attention of engineers. Despite the new sophistication and specialization attending the profession by 1850, most of its members were still simple artisans—carpenters, *metalsmiths, instrument makers*, and millwrights. In France and Germany, they were usually trained in state and military schools rather than in workshops as they were in Great Britain. As a result, continental engineers continued to enjoy a greater professional prestige, even though the greatest number of opportunities for employment were still in England.

In North America, millwrights were the earliest engineers worthy of mention. The first American treatise on mill construction and design was Oliver Evans's, *The Young Millwright and Miller's Guide* (1795), subscribed to by both George Washington and Thomas Jefferson. Although the work dealt primarily in grist mill design, it was the spinning and textile industries that soon completely captured the attention of the millwright. As in England, any common carpenter or *mason* could adopt the title of millwright, and there was little formal training involved. Even so, some millwrights, like Britain's

Millwrights designed and repaired the mills that dotted Western Europe. (By W.H. Pyne, from Picturesque Views of Rural Occupations in Early Nineteenth Century England*)*

Jedediah Strutt, who built highly sophisticated and attractive cotton mills in the early 1800's, received great acclaim for their achievements in the field.

As the 20th century approached, engineering entered the modern stage of its growth. Formal and scientific training became more significant, as engineers were called upon to experiment with materials and to be versed in evermore sophisticated scientific and mechanical processes related to electricity, magnetism, and chemistry. Engineers had to become *scientists* and *mechanics*. Industrial processing became highly refined after the electric motor appeared in 1872, and the steam engine was being applied to increasingly sophisticated uses in both transportation and manufacturing. Engineers were needed in all industries, as they originally had been only in textile manufacturing. They were employed by consumer goods manufacturers and by the metal, tool, and machine industries, which themselves lay at the heart of the factory system. Even agriculture and food processing required the scientific training and knowledge of the modern engineer. German engineers continued to be the most highly respected in the profession right up to World War II, owning largely to the superiority of their technical and scientific training.

Chemical engineering has become extremely important in today's world. The development of acids,

Designer-engineer-inventor Alexander Graham Bell displays his tetrahedral cells, which combined ease of construction with great strength. (Library of Congress, 1907)

plastics, pharmaceuticals, and petroleum products has prompted great interest in this field. *Chemical engineers* work to improve the quality and increase the production of these products, with utmost regard to the cost-effectiveness of such procedures. While they typically work in laboratories, the effects of their quiet work are directly felt in the home, where the products they work on are used in great quantity.

Aeronautical engineers develop designs for aircraft as well as ballistic and guided missiles. *Industrial engineers*, a special class of *mechanical engineers*, work closely with factory construction and operation. They analyze production methods, purchase materials, and report on cost-effectiveness reviews, wage-incentive plans, and management-labor relations, as well as designing the physical plant. *Mining engineers* today are generally much more concerned with petroleum and natural gas than with coal. They also study metallurgical processes to find better and more efficient ways of producing metals from ores. *Nuclear engineers* are concerned with the safe development of nuclear power.

Any of these different types of engineers may be involved in their field at various levels: construction, design, development, planning, production, research, implementation, service, or testing. Some engineers specialize in a particular function. *Design engineers*, for instance, design machinery and equipment, and detail specifications, construction techniques, and materials to be used in implementing them. They may even assist in the construction of prototypes. *Production engineers* plan and coordinate manufacturing systems. *Time-study engineers* are efficiency experts. They measure work efficiency and production methods. *Drafters* or *draftsmen*

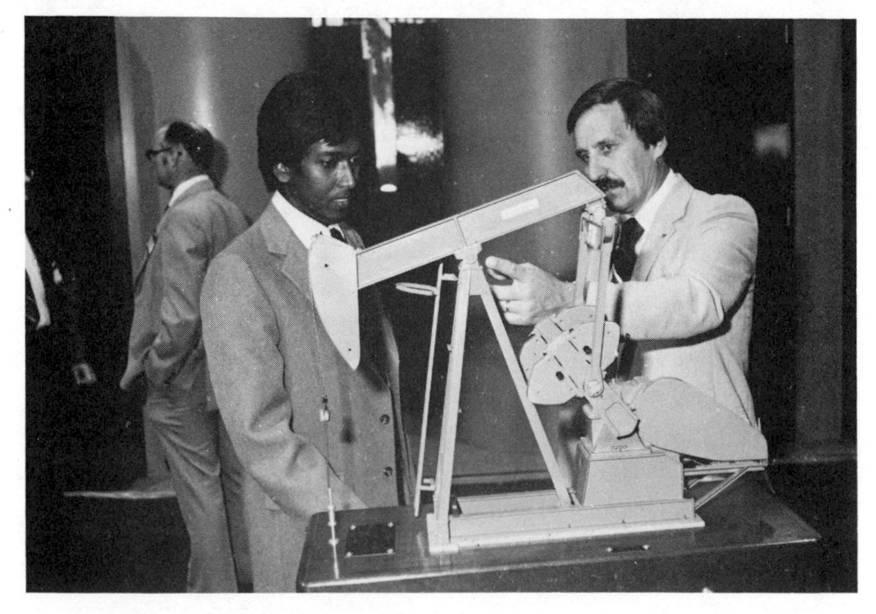

Engineering firms that produce industrial equipment often display it at professional meetings. (Offshore Technology Conference, Houston, Texas, 1981, Society of Petroleum Engineers of AIME)

Electrical and electronics engineers have made possible the communications network on which modern life depends. (The Institute of Electrical and Electronic Engineers, Inc.)

derive detailed construction or production diagrams, drawings, maps, and the like. These are essential in the planning of engineering tasks of every variety. Both *civil* and *mechanical engineers* typically work on projects so extensive and costly that they may be undertaken only by governments or very large corporations. For this reason these engineers are often employed as *civil servants*, or as highly prestigious industrial consultants.

Sudden and phenomenal growth has come in electrical engineering, one of the most significant divisions in the profession today. *Electrical engineers*, virtually nonexistent before the 1920's, now find themselves in the center of the computer and electronic device industries. In addition, they may find employment in other communications and transportation companies. They also deal with large public companies for power generation and transmission, such as that required for industrial and domestic heating, cooling, and lighting. Electrical engineers, like their colleagues in other engineering specialties, must be highly trained in both science and mathematics. They are especially concerned with the practical and utilitarian applications of these principles, theories, and constructs.

For related occupations in this volume, *Scientists and Technologists*, see the following:
Chemists
Computer Scientists
Mathematicians
Physicists

For related occupations in other volumes of the series, see the following:
in *Artists and Artisans*:
Painters
Sculptors
in *Builders*:
Architects and Contractors
Carpenters
Masons
Plumbers
Roadbuilders
Tunnelers and Blasters
in *Clothiers*:
Spinners
Weavers
in *Financiers and Traders*:
Merchants and Shopkeepers
Stewards and Supervisors
in *Harvesters*:
Farmers
in *Helpers and Aides*:
Sanitation Workers
in *Manufacturers and Miners* (forthcoming):
Factory Workers
Mechanics and Repairers
Metalsmiths

Miners and Quarriers
Power and Fuel Merchants
Vehicle Makers
Weapon Makers
Well-Diggers and Drillers

in *Restaurateurs and Innkeepers* (forthcoming):
Bakers and Millers

in Scholors and Priests (forthcoming):
Monks and Nuns
Priests

in *Warriors and Adventurers*:
Sailors
Soldiers

Geographers

Geographers study the physical characteristics of the Earth's surface. Beyond this they study interactions among such physical characteristics and human, economic, and political factors. Many geographers study only certain areas of the Earth, while others apply systematic analyses to geographical phenomena (things or events) such as climate, which is not purely regional by nature. Unlike *historians*, geographers are concerned with space and the physical Earth, rather than with time and society.

The Egyptians studied different regions in northern Africa during their military campaigns, and the Phoenicians investigated the Mediterranean world while engaged in their lively commercial ventures. But it was the Greeks who first created some semblance of formal study and organization of the findings of *travelers*, *warriors, merchants,* and *explorers.* Homer's epic poems suggest that little was known of the world outside of the immediate Mediterranean lands. The Greek navigator Pytheas was the first to fix the position of the lands he had visited, which included Britain and portions of Europe's northern seas. The military expeditions of Alexander the Great went as far as the Persian Gulf and parts of India in the late fourth century B.C.

Greek geographers were usually found among *philosophers* and *historians.* They posed questions concerning the size and shape of the Earth, and ventured some crude answers. Homer thought the Earth was a disk surrounded by the River of Ocean; Thales of Miletus believed it was a floating disk; the Pythagoreans and Aristotle agreed that it was probably round. In the third century B.C., Eratosthenes and Posidonius estimated the circumference of the Earth; though they underestimated it, the fact that they even attempted such calculations showed an increasing vigor in the science. A century later, Hipparchus used mathematical models to divide the Earth into 360 degrees traversed by latitudinal sections called *Klimata.* He was the originator of the latitude-longitude grid pattern used to section the globe. In the second century A.D. Ptolemy summarized the works of earlier geographers, including the cultural and regional assessments made by historians such as Herodotus and Polybius. Ptolemy's work was considered the authority in the field for many centuries, rooted as it was in rather sophisticated mathematical analysis. Strabo, a century earlier, had emphasized the historical study of humankind and cultures as they were related to various geographical areas.

The Romans were more practical than philosophical. They set about making road maps and itineraries for such purposes as military campaigns and postal routes. However, after the fall of the Empire, and for some centuries to come, geographical knowledge and curiosity largely died out in the Western world. The works of Ptolemy were preserved and embellished by the

Moslems, particularly between the ninth and tenth centuries A.D., a period that may be called the golden age of Islamic geography. Ptolemy had always left regional considerations to a special group of geographers known as *chorographers*. He preferred to deal with the world as a whole, and on its largest scale. The Moslems filled in his broad schemes with a great deal of local color from Eastern regions traveled by sugar and spice *traders*. Moreover, they improved on Ptolemy's geography by working out more precise calculations and by developing new instruments for measurements. In the Islamic world, several geographers concerned themselves with the human side of geography, as Strabo had done in ancient times. Ibn Khaldun in the late 14th century is particularly noteworthy, for he established considerable accuracy and detail in the assessment of the interrelationship between people's views and the physical regions in which they lived.

Meanwhile, in Europe there were virtually no geographers, outside of a few *monks* who were primarily concerned with theological matters. In the sixth century A.D., for example, one Christian theologian used vague Biblical references to "prove" that the Earth was flat and rectangular, with Jerusalem at its center, surrounded by water, and lighted by a sun that hid behind the mountains at night. There seems to have been little interest in understanding geography for itself, but only as a symbol of church doctrine and belief or for more mundane purposes, like plotting road maps or sailing charts. Nonetheless, geographic information continued to accumulate through the travels of merchants and armies. Glaciers and geysers were described in a 13th-century Norwegian report, and early in the 14th century Marco Polo's accounts of Eastern cultures began to spread through the West.

As early as the 13th century, the English scholar Roger Bacon had updated the attempt to mathematically calculate the size of the Earth; his insistence that the Earth was round and could be circumnavigated stirred a new interest in geography. But his views were too revolutionary for the times. Meanwhile, the traditional authority of Ptolemy's ancient *Geography* prevented the expression of several new and helpful ideas and findings in physical geography; they were to be overlooked for several centuries to come.

It was only during the Age of Discovery, in the 15th and early 16th centuries, that the shortcomings of Ptolemy's geographical models finally came to be recognized, particularly by *navigators* and explorers who had been misled by them. Columbus's voyages proved to Europeans the existence of whole continents that neither Aristotle nor Ptolemy had known anything about. Columbus had tried to persuade the Spanish monarchs, Ferdinand and Isabella, that Bacon had been correct in postulating that the Earth could indeed be circumnavigated. But it was not until 1522 that Ferdinand Magellan's expedition actually demonstrated the validity of that theory, finally exposing the

Cosmographers like this one developed a modern view of the world, based on accounts by explorers. (Engraving by Stradanus, early 16th century, National Maritime Museum, Greenwich)

vulnerability of the Ptolemaic model, which had envisioned seven separate seas rather than one continuous body of water ultimately connecting all lands.

Vast improvements were made in cartography in the 16th century, upgrading the general knowledge of the physical Earth. Gerardus Mercator initiated the use of a *cylindrical projection* to depict accurately the rounded Earth on a two-dimensional map. Now known as *Mercator projections*, this was but one of the innovations that a series of Flemish *cartographers* used to improve the geographer's perspective of natural boundaries and the relationships of lands and seas. Mercator also drew up collections of maps on specific areas in Europe. When published as a book, they were referred to as an *atlas* because of a picture of the Greek god Atlas on the cover. Thus, Mercator not only devised a new cartographic device, but also more generally rekindled interest in *areal geography*, the study of specific regions, a study which Ptolemy had left to the chorographers.

This work, in many ways, opened the door to the new era of modern geography, particularly when coupled with the precise adaptation of Jean Picard's mathematics, which greatly improved precision of calculation and made triangulation a key model for quantification studies. In triangulation, the location of a certain point on Earth is determined by its relation to two other points, one on Earth and one in space; the angles formed by the imaginary lines joining these points allow the unknown point to be located precisely. Picard's use of triangulation was particularly evocative, since it was based on astronomical and telescopic observation, thereby extending the awareness that geographical data were intimately related to the entire universe.

Although Eratosthenes had been the first to use the term *geography*, he generally regarded it as only the study of the physical characteristics of the Earth. By contrast the geographers Strabo and Herodotus in the ancient world, and Ibn Khaldun among the Moslems, had considered the relationships among peoples, cultures, and lands. Modern geography of the 16th century saw a stricter separation between the special investigations of chorographers and the broader views of geographers. Petrus Apianus, with his mathematically based charts and coordinates, would have been considered a prototype of the geographer, while Sebastian Munster's descriptive and local studies were ideally in the realm of chorography. Perhaps nobody more than Bernhardus Varenius, in the 17th century, opened geographers' minds to fresh examinations of the Earth and its relationships with people, when he divided the field into *general geography* and *special geography*—the latter for the most part concerning chorography. This so-called *Varenian framework* gave geographers a clear call for searching out the relationships between human populations and specific areas. Unfortunately, Varenius's early death at age 28 prevented him from developing more precise methodologies for special geography, and this enterprise was not vigorously approached for quite some time to come.

The 17th and 18th centuries saw a steadily increasing accumulation of physical data, owing largely to a new emphasis on rationalism followed by a revolution in instrumentation. The Copernican or Scientific Revolution of the 16th century was based on new evidence supporting the contention that the Earth was in constant motion along an orbital path that circled the Sun. For centuries, the Ptolemaic model of a stationary Earth in the center of the universe had been accepted without question. In fact, when questions did arise, the curious were called heretics and infidels, since the church had totally endorsed Ptolemy's views. Once those long-revered astronomical models were shattered, however, it left scientists in other fields wondering about ideas that they had always blindly accepted. If Copernicus and Galileo could use nothing more than reason to prove Ptolemy's universal model untrue, after it had been the sole authority on the subject for well over a thousand years, what else could be discovered through the application of a rational *scientific method* to other fields of inquiry? Geographers were among those new scientists who first overcame old, unproven theories—many of which, ironically, had also been handed down from Ptolemy. They set about to observe and test what was actually the case. Geographers, like other scientists after 1600, began to divorce themselves from philosophy and theology. They delve into observable data, from which a scientific body of knowledge and information could be built.

Greatly enhancing the new desire for hard facts was the development of the printing press and the invention of many new instruments, particularly those used in *geodesy*—the study of the size, shape, and even internal structure of the Earth. The printing press helped popularize the new scientific methodologies, as well as making banks of information and contrasting theories widely available to the inquisitive and pioneering geographer. Much of the new information that became available for publication was obtained through geodesy. An ancient aspect of geography, geodesy had entered an exciting new phase with Picard's use of triangulation and computational logarithm tables, which became significant tools in measurement. Thermometers, barometers, chronometers, and hygrometers all came into use for working geographers as well as for navigators and explorers.

Some geographers concentrated on refining mathematical computations to better illustrate the condition of the Earth's surface. However, most of the research during this period was in the field of physical geography, with a stress on collecting data on a wide variety of natural phenomena. The British *astronomer* Edmund Halley was the first to produce a wind chart and a theory correlating the trade winds with the Earth's heat. Torbern Bergman was a pioneer in producing *gravimetric determinations*. He posed the existence of affinities between certain minerals found in the earth, and employed quantitative analysis in determining the precise nature of specific mineral compositions. He did

In modern times, geography became an important part of the standard school curriculum. (From The Good Things of Life, *1886)*

this by carefully weighing chemical precipitates of minerals, to determine their specific gravity. In the 18th century, Philippe Buache, the first geographer to make use of contour lines, stressed individual physical phenomena in his published works. These summarized a good deal of the new data being gathered during this period. He and others also began searching for underlying connections between the great mountain and basin systems.

All this activity may be summarized by the popular term *pure geography*, as it was then thought of. It must be noted, though, that even the most prolific and original researchers in the field confined their studies to largely artificial political boundaries, which really had little to do with geographical phenomena. For example, Jedidiah Morse, the father of American geography, divided information that he had gathered according to individual states in his *Geography Made Easy*, published in 1784. Half a century earlier, Leyser had looked into the matter of natural boundaries as a foundation for investigation. As far back as 1625, Nathaniel Carpenter had focused his attention on the areal relationships of physical things, and Philipp Cluverius had already written detailed regional descriptions. But it was not until the early 1800's that this idea of natural regions, as opposed to political regions, gained the proper attention of geographers. Nonetheless, the profession had become active and visible enough, during its *pure* period of collecting data, for it to become somewhat better recognized as a unique occupation, and not one strictly married to cartography and navigation. For several years in the second half of the 18th century the famous German philosopher Immanuel Kant, even lectured at the University of Königsberg on the subject of physical geography.

Somewhat inspired by the published account of Kant's lectures, which gave geographical studies a prominent place in the history of logic, Alexander von Humboldt, along with Carl Ritter, opened up a new age in geography beginning in the first half of the-19th century. Although Humboldt had a broad overview of the Earth's features—his five-volume work *Kosmos* was the first decent encyclopedia of geography and geology—he emphasized the terrestrial harmony of a given region. He always sought underlying correlations between specific phenomena and general laws, and was the first to use *isothermal lines*— which mark the equal temperatures of different places at the same time—in discerning climatic zones. Ritter's interests were much broader than physical geography. He saw the science as it is related to the human condition and even the progress of history. He correlated physical and cultural features of specific regions, and thus stimulated interest in areal geography. Still, he insisted that such studies ought to be comparative and not generalized to the point where one might assume that the features of one area are the same for all others or even all other similar areas. The works of Humboldt and Ritter together gave geographers a whole new outlook on their discipline, and gave the profession greater scientific and academic prestige than it had ever had.

In the 19th century Humboldt was generally thought to have preferred a *general*—or *systematic*, as it is now called—investigation of geography,while Ritter was supposed to have championed regional, or *areal*, studies. Modern examination of the writings of both men reveals, however, that their approaches to their discipline was far less simplistic than that, and both seem to have recognized the interdependency of the two modes. The main influence that they had was to make geography a true science. The remainder of the 19th century saw a great many more professionals in the field, with most of them developing a tradition of direct observation and careful calculation rather than vague philosophies based purely on logic, religion, or otherwise.

Chorographic studies (concerning areal differentiation and interrelationships) came to be more purposeful and less encyclopedic as geographers in the second half of the 19th century began studying specific regions to uncover general themes or rules, or to investigate predetermined relationships. The goal became focused on understanding the causes and consequences of areal differentiation, rather than studying a large collection of unrelated physical and cultural phenomena. Following Humboldt's and Ritter's lead, two other German geographers, Ferdinand von Richthofen and Alfred Hettner, further defined the course that modern geographers would ultimately follow. Working in the late 19th and early 20th centuries, Hettner believed that geography was chiefly an areal science of the Earth's surface, but one that sought comparison and synthesis of models, not mere descriptions and abstractions of them. Richthofen, at the same time, made the point that chorography might entail little more than a "nonexplanatory description" of physical regions, but that such data ought then to be synthesized by systematic explanation of causes and consequences. This *areal synthesis* should be the occupation of geographers concerned with *chorology*, as the completion and elaboration of *chorography*.

John Pinkerton had written in 1807 that geography "only aspires to illustrate history" and, indeed, special geography—the aimless collection of odd facts—was

These surveyors helped plan the route of the Deadwood Central Railroad, while their kin were surveying and mapping the West. (Library of Congress, 1888)

quite popular at the time. Meanwhile, the teaching of general geography continued to accompany the more serious and practical studies of cartography and navigation. By the beginning of the 20th century, however, geographers were considered to be quite sophisticated in scientific and scholarly research. Practitioners of physical geography had produced detailed studies of the formation of mountains, seas, rivers, and continents, while the role of rivers in giving the Earth its present shape had also been explored. William Morris Davis studied the cycle of erosion, and historical approaches opened up questions about stages of the Earth's development. Systematic (general) geography, too, had largely focused on physical geography. Climatology expanded with the use of mean monthly isotherms (lines connecting places with the same average monthly temperatures) and Berghaus's use of *isohyets* (lines connecting places with equal rainfall at the same time). *Meteorology*, too, matured as patterns of wind, weather, and even prehistoric climates were thoroughly researched.

Many special branches of geography arose. Mathew Maury's publication of the *Physical Geography of the Sea* in 1855 marked the real beginning of modern *oceanography. Animal geography* was enhanced by the studies of Charles Darwin, including his theory of evolution. *Plant geography, soil geography,* and *biogeography* are all concerned with different aspects of the Earth's surface. *Human geography* had never been terribly fruitful, since most geographers preferred to study only the physical features of the earth (*pure geography*). This bias began to change with Darwin's publication of the *Origin of Species* in the middle of the 19th century. By the end of the century, Friedrich Ratzel—using Darwin's basic concepts—began studies of the influence of nature on human beings. He also gave some thought to the influence of people on nature, but this view was little considered until the work of Paul Vidal de la Blache somewhat later.

Vidal's influence on contemporary geography has been enormous. His main thesis was that nature offers choices to human beings, but they make the choices of how they will deal with nature and the physical environment. He completely rejected Ratzel's notion of *environmental determinism*, that is, the idea that people have different natures, depending on the type of geographic area in which they live. Vidal insisted that the *genres de vie* (lifestyles) of cultures and peoples left an impress upon geographic regions much more than people bore the impression of the physical features of a region. His studies were mostly regional and areal and had great influence on the French tradition in geography.

At this same time, Ellen Churchill Semple and Ellsworth Huntington were giving American geography an image of environmental determinism. Semple was concerned with the influence of landforms on American history, while Huntington spoke of climatological determinism in his *Pulse of Asia* in 1907. The British, too, lagged behind the advanced thinking of Vidal and his French movement. Despite such ideological differences within the profession, the general esteem for geographers throughout the Western world improved vastly, beginning with the middle of the 19th century. The Germans had always held a high regard for the professions; Carl Ritter was appointed to the first Chair of Geography at the University of Berlin in 1820. Later the Royal Geographical Society in England had several members who attained university posts, foremost being Halford J. Mackinder, who was appointed to Oxford University in 1887. Mackinder was a brilliant lecturer who introduced the British to the benefit of regional studies like the ones Vidal had undertaken.

In the United States, the American Geographical Society was founded in 1852. Geographers as a group were poorly trained in America before mid-century, when the Swiss-born, German-trained Arnold Henry Guyot introduced the instruction of scientific geography at Cambridge, Massachusetts. In 1854 he became professor of physical geography and geology at Princeton University. After the Civil War, George Perkins Marsh pioneered the field of conservation, inspiring later studies of erosion cycles. One of America's most prominent geographers was naval officer Mathew Maury.

The 20th century has seen considerable interest in human and environmental geography. The "geographic factor" has most recently been regarded in fuller perspective as only one factor among many related to *human*

Early meteorologists sometimes ascended in balloons to gather data on weather patterns. (From Voyages Aeriennes, *by James Glaisher, 1870)*

ecology. Geographers today are much more inclined to study mutual effects of living beings and nature upon each other. But in order to study intricate relationships, *ecologists* needed improved methodologies, instrumentation, and calculation. Moreover, it became much more manageable—given the larger number of variables being tested for in controlled or synthetic studies—to look at smaller areas in order to more fully view broader concepts. In effect, geographers began to look at the subjects of their investigations, as biologists would study a piece of tissue under the microscope, seeking the most minute and subtle components and configurations, which might eventually unlock the most profound questions of life itself. The geographer today makes wide use of almost microscopic areal studies, from which systematic hypotheses may eventually result. Hence, there are studies of climatic regions, agricultural regions, industrial regions, economic, political, urban, and rural regions, to mention just a few of the broad topics surveyed in areal geography.

Today, most professional geographers agree that their research ought to be directed toward specifically stated ends. This has meant that there is more direct field observation involved in the occupation, particularly in topographic studies of restricted scope, such as studies zeroing in on specific neighborhoods, fields, or even city buildings. As early as 1915 two American geographers, W.D. Jones and Carl Sauer, published a detailed field study of one specific agricultural area that treated all relevant features in explicit detail. This sort of study is now typical of the way geographers in all areas approach their inquiries.

Quantification methods have particularly added to the precision and predictability of modern geography, particularly since 1950. Statistical analysis and advanced computerization have permitted geographers to use a wide range of field studies and specific data in order to engage in predictive model building and systems analysis related to both areal and systematic studies. Advances in technology have led geographers to once again explore general laws and principles which underlie all spatial and ecological patterns. But unlike earlier systematic studies, which were based on philosophy and pure theory, contemporary studies are rooted in more specific data and observable case studies.

A great many geographers with university degrees are employed today as *teachers*. Many more work for federal, state, and local government agencies in the regional and urban planning services. Still others find work in marketing, advertising research, publishing, and industry as consultants. Skills required in most of these professions include the analysis of aerial and ground-level photographs, mapping and computer programming, cartographic methodology, and recording of direct observations through written outlines and field sketches. The education and training of geographers may cover virtually all areas of research, calling forth an endless variety of fields of specialization. This is because geography is really the study of the spatial variations and relationships of everything on the Earth's surface, and even to some extent the interior. It involves the areal and systematic study of rocks, rainfall, climate, people, politics, economy, plants, animals, and anything else that goes into the "shaping" of a region. Generally speaking, most of these things and many others are intricately interrelated in their effects on any given geographical "situation" encountered.

Closely related specialists, *climatologists*, study comparative climates of various regions or world sectors, as well as the details of the specific climate of any given region. The ancient Greeks differentiated between *regional climatology*, which accounted for the typical average climate of a region, and *physical climatology*, which accounted for actual day-to-day atmospheric conditions. With the advent of *meteorology* as a specialization of its own, by the mid 19th century at the latest, climatologists turned most of their attention to the study of regional climates. The most recent trend in the field has been to study comparative climates in their dynamic relation to one another.

Meteorologists study weather conditions, patterns, and what determines them. They are specifically concerned with the relationships among atmospheric elements such as pressure, wind, and moisture. There was no scientific investigation of the weather before the 17th century, and no real professional development in the field until the 19th century. At that time, new instruments such as barometers, thermometers, and hygrometers began to bring the art of predicting the weather beyond the levels of pure intuition or mystical insight.

Weather diaries were being recorded by the 18th century, and several institutions were established for the study and recording of such information. Meteorology as a professional service began in the 19th century and was closely related to military and merchant marine operations as well as agricultural planning. The first weather maps were drawn up in the middle of the same century.

Since World War II, meteorologists have provided their services to mass audiences via television, radio, and newspapers as well as old-fashioned almanacs. Today, observations of the atmosphere from space have allowed scientists in this field to study the dynamic and constantly changing atmospheric conditions and relationships that determine the weather. In this way they are able to follow developing and dissipating weather patterns well in advance of their actual appearance in a given location.

For related occupations in this volume, *Scientists and Technologists*, see the following:

Astrologers
Astronomers
Cartographers
Chemists
Geologists
Mathematicians
Scientific Instrument Makers
Statisticians

For related occupations in other volumes of the series, see the following:

in *Communicators*:
Authors
Messengers and Couriers

in *Financiers and Traders*
Merchants and Shopkeepers

in *Scholars and Priests* (forthcoming):

Monks and Nuns
Scholars
Teachers
in *Warriors and Adventurers*:
Sailors
Soldiers

Geologists

Geologists are concerned with a specific aspect of geography, namely systematic and areal studies of the materials—mostly rocks and minerals—that form the surface of the Earth's crust. In their concern for the Earth's surface, geologists usually must adapt principles of physics, chemistry, mathematics, and biology to analyze their subject. Moreover, they are frequently called upon to review their findings with respect to history and economics.

Ancient *philosophers* and *historians* speculated widely about matters of concern to today's geologists. Herodotus, for example, determined that "Egypt is the gift of the river," based on his observations that the soil of the Nile's fertile plains included heavy amounts of riverbed deposits. Such evidence was usually quite general and speculative, however, and no methods of classification were developed beyond the painstaking attempts of Aristotle.

For the most part, geological data gathered before the 15th century A.D. was disconnected, pointless, and frequently the result of speculation rather than direct observation or scientific reasoning. But Avicenna, the medieval Islamic scholar, made profound contributions to the understanding of geology, particularly in his works on the formation and erosion of mountains as related to the origin of rocks. The Renaissance artist and scholar Leonardo da Vinci began to pose intriguing questions (and propose surprisingly farsighted answers) about the study of the Earth's crust formations. At the same time the word *geology* itself first came into use. It was derived from the Greek word *ge* meaning the *Earth* and *logy* meaning *knowledge of*—in contrast to the much higher form of knowledge, *theology*, knowledge of God or divinity. The word *geology* was not commonly used, however, until the science itself came to be regarded as a worthy enterprise in terms of scientific understanding. That was not until the 19th century.

Georgius Agricola in the 16th century wrote *De Re Metallica*, one of the greatest books on *mining geology*, an incisive account of ore veins, rock formations, and ground water, as such information pertained to the mining of precious metals and minerals. His *De Natura Fossilium* was a landmark work in *mineralogy* (the chemistry of minerals). Most *scientists* during this time thought that fossils had been created inside of rocks, or were perhaps the remains of animals destroyed during the Great Flood described in the Old Testament. A Swiss physician named Conrad Gesner was the first to make a solid case for fossils as animal remains deposited throughout history; he published his findings in 1565.

However, the Old Testament versions of Creation and the Flood continued to dominate geological inquiry for quite some time. The 17th-century Irish bishop James Ussher made the famous calculation that the earth was 6,000 years old according to Scripture. He dated Creation at 4004 B.C. and the Flood at 2349 B.C. Geologists were hampered by these figures for nearly two centuries, not daring to contradict them for fear of being labeled infidels or, at best, poor scientists.

Nonetheless, advances were already under way. People such as Robert Boyle made important investigations in mineralogy. Nicolaus Steno and others had already established the idea of stratification through the "law of superimposition"—the geological concept of earth layers (*strata*) being deposited one on top of another in such a way that the newest are those closest to the surface. By the end of the 17th century the English physicist Robert Hooke had proven that fossils were not "sports of Nature," but indeed the remains of previously

Geologists have long worked closely with miners, here laying out the digging of new mines. (From Diderot's Encyclopedia, *late 18th century)*

living organisms. Landscape studies were advanced primarily by Steno, who had discovered the essential role of water systems in shaping the Earth's landforms.

By the 18th century theological or philosophical speculation in the field was being steadily replaced by direct observation and scientific methodologies. Advances in biology, chemistry, and physics had made this inevitable. Gradually, it was becoming clear that the Earth must be much older than thought by *creationists* (those who held firmly to Ussher's calculation of 4004 B.C. as the date of Creation). Geologists began to understand that the long process of geological change had required much more than 6,000 years to develop. The exploration of these ideas culminated in James Hutton's 1785 reading of his *Theory of the Earth* to the Royal Society of Edinburgh. Hutton declared that the Earth's shapes and formations occurred according to the principle of *uniformitarianism* rather than *catastrophism.* That is to say, he demonstrated that the Earth changed and developed gradually through uniform principles rather than through catastrophic events, as had long been believed by creationists and others.

Hutton was also involved in the greatest controversy in geology of his time, that between the *Plutonist* and *Neptunists.* Hutton and the Plutonists held that heat and volcanic action had led to most of the geological change throughout history. The eminent German geologist A.G. Werner led the Neptunist argument that the Earth's strata had been deposited mostly as sediment through the action of water. Werner was a highly regarded teacher and scholar, who operated his own school of mining and geology in Freiberg, Germany. His views of Neptunism held the day against Plutonism. Indeed most of Werner's views held the day against Hutton's, not only because of his greater popularity but also his theories were much more orthodox. Neptunism seemed to bear out the story of the Flood, and others of his teachings fit the church's idea of the true age and chronology of the Earth.

In an age when radical movements (like the French Revolution and its offshoots) were shaking the political foundations of Europe, many scientists were reluctant to entertain unorthodox views. But a Scientific Revolution was in full swing, too. Neither the authority of the church nor that of the state could permanently forestall geology's claim to being counted among the new sciences. Half a century following Hutton's death, he was finally recognized as the father of modern geology. The transition was marked by the publication of Charles Lyell's astoundingly successful and influential *Principles of Geology* in 1830. Based almost entirely on Hutton's work, it was the first sound textbook in the field and went through 12 editions by 1875. Perhaps the most influential of Hutton's ideas was the concept of evolutionary changes, which eventually helped shape Darwin's *Origin of Species* and establish the concept of biological evolution.

By the mid-19th century, geologists had largely escaped their role as *surveyors* of mines. Of course, mining was an extremely lucrative enterprise, and both industry and monarchs had for a long time sought the services of geologists in the working of private and state mines. With the Industrial Revolution upon them, geologists were still demanded by the mining business, especially in opening up and working coalfields. But geology no longer was solely dependent on that industry, nor was it simply an adjunct to *geography*, another role into which it had been typecast since ancient times. *Paleontologists* had shown that geology was a method for the study of both history and biology.

The thorough geological study of all land areas was under way by the mid-19th century and continues today. Better systems for the analysis and classification of rock materials improved the exactness of geological studies—and, consequently, the esteem in which geologists were held. William Smith, the father of *stratigraphy* (the mapping of geological strata), introduced new methods that gave such diagrams much greater accuracy and detail.

The first national geological survey was founded in England in 1835. Additional surveys were soon established in other countries, most of them emphasizing the location of precious mineral deposits. The United States Geological Survey was established in 1879.

Techniques for studying minerals and rocks were improved considerably after the mid-19th century. The invention of the polarizing microscope opened the new fields of *petrology* and *petrography*, the classification of crystalline rocks through a study of characteristic optical properties. These properties could be detected by the telling penetration of light through thin rock slices, which gave readings through the polarization microscope. This represented a mild revolution in the study of minerals, which today has been enhanced even further through the use of X-ray machines. In 1840 Louis Agassiz proposed his *glacial theory*. From his observations of glacial movements and deposits, he concluded that ice caps had once covered extensive regions in northern Europe, Canada, and the United States. The subsequent development of *glacial geology* helped explain the origins of rocks, minerals, and even landforms and landscapes. Further systematic study of landforms followed the explorations of John Wesley Powell and Grove Gilbert in the American West, notably along the Colorado River with its Grand Canyon.

Today, the use of advanced technology has made geology a very sophisticated occupation. X-ray equipment, electron microscopes, and the mass spectrometer are just a few of the instruments that have helped people to better understand the Earth's surface and age—the latter being the specific concern of *geochronology. Oceanography, glacial geology,* and even *astrogeology*—the study of extraterrestrial land surfaces—have become important specializations within the field.

Geologists are employed as *teachers*, oil and mineral *prospectors* and *explorers*, and industrial and government *consultants*. Their studies may be divided into three main categories: physical geology, historical geology and geological mapping. *Physical geology* includes *mineralogy* and *petrology*, studies focusing on the origin and composition of rocks, minerals and the Earth's crust; *geomorphology*, the geological history of surface land forms, and *paleogeomorphology*, the history of ancient, buried land forms; *structural geology*, the study of rock structures and their origins; *hydrogeology*, the study of the nature and movement patterns of groundwater; *geochemistry*, the study of the excavation of the Earth for industrial uses; and *economic geology*, the study of the location and value of those of the Earth's resources that are useful to humans. *Historical geology* is concerned with *stratigraphy*, the analysis of stratigraphic patterns and origins, and with *paleontology*, the study of fossils embedded in sedimentary rock strata. *Geological mapping* is concerned with charting and diagramming the surface land, water, and mineral conditions of specific areas.

Among related specialties, one has proved of special importance in modern times: paleontology. *Pa-*

Many a geologist—amateur and professional—has attempted to take pickaxe in hand and seek a fortune, like this prospector. (By Arthur Rothstein, from The American West in the Thirties, *Dover, 1978, in Nevada, 1940)*

leontologists study fossils, the remains of animals and plants found in sedimentary rocks. Those who study either vertebrate or invertebrate animal fossils are *paleozoologists*, while those who study plant fossils may be called *paleobotanists*. *Paleogeographers* depict the prehistoric world in map and chart form, based on geological and fossil evidence.

Paleontology became an established science with the work of Baron Georges Cuvier in the late 18th and early 19th centuries. He took Carl Linnaeus's systematic classification of living things and adapted it for the first time to classify fossils in zoology. A little later, Adolphe Brongniart pioneered the field of paleobotany, establishing it as a special branch of paleontology. Through comparative classifications of fossil remains, paleontologists were quickly drawn into the evolution controversy. In general, they found that the more ancient a fossil was, the more simple it was. The complexity that appeared gradually in later epochs indicated an actual evolution in biological sophistication—and of entire species.

Today, paleontologists study the age and formation patterns of rocks and their sources, whether they originated on land or in water. In this manner, they can determine the age of land and seas and can make detailed maps of the Earth as it must have appeared during different epochs. Many paleontologists are hired by oil companies to help determine good sites to drill for oil. This sort of scientific prospecting is based largely on careful studies of fossil deposits at a given location. Scientists in this new field work closely with *historians, geologists, geographers,* and industries desiring knowledge of the sources and compositions of specific areas of the Earth and its continental shelf.

Another closely related specialty is oceanography. *Oceanographers* study the marine environment. They are concerned with the waters of the ocean—the marine life in it, and its chemical and mineral properties—as well as the ocean "floor," the solid earth beneath the waters.

Much of their work involves the mapping of oceans and their surface or subsurface landmasses.

People have always been concerned with the ocean as a matter of geographical, commercial, and cartographical investigation. The quest for this knowledge increased considerably during the Age of Discovery and with the finding of the New World. It was not until the 18th century, however, that Captain James Cook set out on a scientific expedition intended to produce information about the geographical features and measurements related to the ocean.

While the practical aspects of oceanography continued to be pursued into the 19th century, the new spirit of science finally led to investigations of the very nature and properties of the ocean itself. The Danish geologist Johann Forchhammer had chemically analyzed surface samples of seawater for some 20 years before the first truly scientific expedition explored the ocean depths. The British *Challenger* sailed between 1872 and 1876, amassing such information and thus marking the modern foundation of the occupation. This investigation resulted in the presentation to the Royal Society of a 29,500-page report that had enlisted the expertise of some 76 authors and scientists. Other expeditions followed this example. The Norwegian *Fram* (1893-1896) investigated the Arctic Ocean, and the British *Discovery* (1925-1927) and *Discovery II* (1932-1934) explored the Antarctic waters.

After World War I, oceanographic institutions sprang up to systematize the scientific searches for new data. Better instruments and improved oceangoing vessels contributed enormously to the maturity and specialization of the profession. Today, oceanographers make detailed maps of the ocean and its bottom. They search for proper drilling sites for oil and gas, explore the ocean's other potential energy sources (tidal, wave, and heat), and work at locating and developing both food and mineral resources from the sea. They are employed by government and industry and hold posts in most major universities. With the widespread testing and acceptance of the Continental Drift Theory—the idea that the Earth's continents have, over eons, moved around the globe on a crust no longer recognized as static—oceanographers and geologists have been working ever more closely together in the field.

For related occupations in this volume, *Scientists and Technologists*, see the following:
Biologists
Cartographers
Chemists
Geographers
Mathematicians
Physicists

For related occupations in other volumes of the series, see the following:
in *Manufacturers and Miners* (forthcoming):
Miners and Quarriers
in *Scholars and Priests* (forthcoming):
Scholars
Teachers
in *Warriors and Adventurers*:
Sailors

Mathematicians

Mathematicians are scientists who work primarily with the language of numbers. They study the world around them and express the patterns and relationships they see in numerical form. Since ancient times, mathematicians have provided the essential tools for analysis and measurement in all the other sciences.

Even before the establishment of the ancient civilizations of Egypt, Mesopotamia, India, and China, people used crude measurements in devising basic techniques of building, cooking, hunting, and craftwork. But it was the growth of the great early civilizations that give rise to the actual study of mathematics. *Geometry*—literally, "measurement of the Earth"—was applied in antiquity to define land boundaries and ownership rights over specified lots. Royal *surveyors* were trained by court mathematicians, as were *tax assessors, engineers,* and *astrologists*. Their jobs were considered extremely important by the rulers, who depended on accurate records regarding taxation as well as payments, usually non-monetary, made to military and other personnel. The large staffs of *scribes* that helped hold together the administration of many a great empire were drilled in practical uses of numbers, measurements, and the keeping of accounts.

Mathematicians were instrumental in the most significant aspects of ancient life. They worked closely with, and often doubled as, *astrologers*, who had an exalted role in society, especially at the royal courts. They conceived, organized, and often supervised engineering projects, such as draining marshlands, irrigating deserts, and controlling the floods of the great rivers: the Nile, Tigris, Euphrates, Indus, and Ganges. They also handled building projects such as temples, pyramids, and other public edifices. Some mathematicians were also *teachers*. Many a great and prosperous businessman would gladly foot the tutoring fee of a respectable mathematician to ensure that his son, who would one day take over the family enterprise, was well versed in the use of arithmetical operations and accounting techniques. The curriculum rarely included abstract understanding, however, which most such tutors themselves lacked.

Mathematics emerged from its close association with arithmetic and practical applications when the Greeks began to probe its deeper dimensions. Greek mathematicians were the first to develop abstract principles in their field, and to truly consider the nature of numbers and geometric relations. They also began the practice of grounding mathematical and geometric theory in rational logic rather than natural analogy. That is, they began to recognize the usefulness of designing mathematical models that could help solve problems in the natural and real world by virtue of their being able to be explained coherently through logical deduction. These models—even though purely manufactured by mathematicians—could then be generalized to include real situations in nature or otherwise. In the East, mathematics was also being developed somewhat beyond mere trial and error methods at this time, but the grounding of mathematics in abstraction and logic belongs solely to the Greeks.

Thales of Miletus was one of the pioneers in applying *deductive reasoning* to geometry, while Pythagoras and his followers were among the first to establish *axiomatic procedures* to establish the mathematical proof of a proposition. Axiomatic procedures are those whereby propositions may be directly or indirectly deduced—that is, logically demonstrated—from self-evident postulates (statements or axioms) made about a particular set of mathematical objects or symbols. Through a related series of axioms concerning the relationship among the sides of a right triangle, for example, the famous *Pythagorean theorem* was deduced. It stated this: The square of the length of the hypotenuse (the side opposite the right angle) is equal to the sum of the squares of the lengths of the other two sides. The real significance of the Pythagorean theorem was in demonstrating how propositions and theorems could be deduced logically from the axioms of a given statement of terms. Mathematics became firmly rooted in reason.

Two and a half centuries later, Euclid, one of the most famous mathematicians ever to have lived, compiled the geometrical knowledge of the Greeks up to that time. The resulting textbook, called *Elements*, was the most authoritative work in the field until the 19th century and is still widely used. Other advances were made during the Greco-Roman period as well. In the third century B.C., Archimedes used methods of logic that were not unlike those which eventually developed into *calculus* in the 18th century A.D. In the second century A.D., Ptolemy helped to develop *trigonometry*. This field had been anticipated as early as the sixth century B.C. by Thales, who calculated the height of a pyramid by comparing the length of its shadow to that of a stick that could be directly measured. In the third century A.D., Diophantus made extensive use of equations to determine equal relationships from which deviations could be quantified. He is considered by many to have been the father of *algebra.* Many Greek mathematicians were, like other scholars, attached to academies, the greatest of which was in Alexandria. Though most were men, some few—as would be true throughout history—were women. The earliest woman mathematician we know of was Hypatia, who learned mathematics, including her special field of Diophantine algebra, from her father, Theon, also an eminent mathematician.

Most of what the Greeks accomplished set the foundation for studies that developed in fields related to mathematics. They were quite obsessed with geometry, partly because of their passion for astrology, which made extensive use of geometric forms. Most of the mathematicians of the Classical period and later were *philosophers*, a fact that greatly influenced the Greeks' perpetual quest for the most abstract understanding of mathematical objects and principles. There were really rather few professional mathematicians, except for the handful of teachers of bookkeeping and arithmetic who found employment in schools or else made the rounds on the tutoring circuit. The Romans were much more interested in the application of this new body of knowledge. They used it for administrative purposes—bookkeeping, engineering, and even military science, one of their favorites.

But the Greeks had set the course for other mathematicians to follow for many centuries. The theorems and principles they derived—such as "the whole is equal to the sum of its parts" or "a straight line is the shortest distance between two points"—were held as absolute truths by generations of mathematicians. Mathematics was held spellbound by the idea that Greek knowledge on the subject, as summarized by Euclid, was pretty much final as far as it went. The basic premises of Euclidean geometry were rarely questioned, nor were alternative perspectives offered. Aristotle had realized that laws derived through the reasoning capabilities of man could not be absolute. He commented on the deductive process:

> It is not everything that can be proved, otherwise the chain of proof would be endless. You must begin somewhere, and you start with things admitted but undemonstrable. These are first principles common to all sciences which are called axioms or common opinions.

Succeeding generations of mathematicians had a difficult time in recognizing the general relativity of so seemingly "exact" a science. Casual observers, students, and even scientists often fail to recognize the significance of Aristotle's remarks even today. Mathematics did not begin to realize it in concrete terms until the 19th century. There was, of course, room for expansion in mathematics; the Greeks had concentrated on geometry, but the realms of algebra and trigonometry had only barely been opened when the Roman Empire collapsed. A few pioneers would lead the occupation in new directions over the centuries.

After the fall of Rome, European interest in mathematics was generally confined to a limited number of *monks*, who usually had purely practical goals in studying it. The only professionals in the field were the occasional *tutors*, who taught practical arithmetic, as their earlier Greek counterparts had done. Even this function was largely performed by *priests*, for in medieval Europe they comprised most of the *schoolmasters* and *tutors*. Occasionally *physicians* and *alchemists* would do some investigative work in the field, but they were hardly full-time or professional mathematicians.

Meanwhile, mathematics was making great strides in the East. China, though it clung to a cumbersome (if beautiful) form of script for words, may have first developed the decimal system of place values and the concept of zero that are at the heart of all our modern mathematics. Mathematicians in China were *scholars*, often working closely with *astronomers* on calendrical and astrological calculations for the royal court. Some worked as tutors to Chinese princes, and many developed practical uses for mathematics, which were applied to the complex task of administering the empire. Many of these mathematicians were monks from among the various Chinese religions; it seems likely that Buddhist monks on pilgrimages were responsible for bringing some Chinese mathematical innovations to India.

Mathematics found fertile ground in India. Indeed, scholars long thought (and some still do) that India itself developed the place-value system, which was in use there by at least the sixth century A.D. at latest. Indian mathematicians probably developed the practice of writing the symbol for zero. This made possible the contributions that are indisputably Indian: notably the development of basic algebra and a simplified, easy-to-manipulate numerical system. While Europe was in its Dark Ages, Indian mathematicians such as Brahmagupta, Mahavira, and Bhaskara were developing means of extracting square and cube roots, and calculating a more accurate value of "pi". Like mathematicians elsewhere, they worked closely with astronomers, making practical contributions in trigonometry and geometry, and even taking some steps toward calculus. Indian mathematicians (who also learned from the Greeks) were highly regarded in Asia: The Arabs to their west actually called mathematics *hindisat*—the Indian art.

It was to the Arabs that the torch passed next, as mathematicians were gathered along with other scholars at the royal courts in the golden age of Islam. The Arabs and Jews who worked in the Moslem courts continued the abstract mathematical studies pioneered by the Indians, especially in the area of algebra. Equally important, these scholars preserved many of the outstanding Greek mathematical works and eventually passed them on to the Western world, which began translating them into Latin as early as the 12th century.

The scholars of the Moslem world did not simply pass

on translations of Greek mathematical treatises, but tremendously expanded those ideas. Particularly significant was the central role of astronomy in Islamic mathematical computation. Indeed, astronomy proved to be the greatest catalyst to serious mathematical study in both the East and West. Greek mathematical astronomy had found its greatest champion in Eudoxus, who described the apparent motion of the planets through an elaborate system of spherical links. The Moslems contributed to this research, particularly through their explorations into plane and spherical trigonometry; these studies culminated in the works of Nasir al-Din al-Tusi, who used all six trigonometrical functions and described the spherical triangle in the 13th century A.D. The pioneer studies of al-Khwarizmi in the ninth century and Omar Khayyam, who classified cubic equations in the 12th century, put algebra on a firm scientific footing by the time the Europeans began translating their research. The medieval works of Nasir al-Din al-Tusi and ibn-Yunus even anticipated the eventual developments of, respectively, non-Euclidean geometry and logarithmic calculation many centuries later.

In Europe, the occupation of mathematician began to revive with the university movement that started slowly in the 13th century. Several noted scholars became *professors* and *lecturers* in mathematics, although it was not really until the Renaissance that the subject first stood on its own merits. Even then, that was due largely to the influence of the Copernican Revolution, which began with the startling revelation that the Earth was not only in motion, but also orbited the Sun in a circle. For countless centuries it had been assumed that the Earth was the center of the universe, a view held tenaciously by the church, which viewed human beings as central in God's creation. To discredit such a firmly ingrained view, Nicolaus Copernicus used neither theological disputation nor divine revelation, the common sources of authority before the 16th century. Instead, he appealed to people's intellectual abilities to deduce planetary motions and positions through observation and the application of mathematical concepts to the physical nature of the universe. To many, this method of cold calculation and measurement reduced the universe to a great machine and God to a *deus ex machina* (god from a machine). But to those who saw the usefulness of such rational approaches to knowledge, the importance of mathematics to uncovering the physical laws of the universe became clear.

As a result, mathematics—which had previously been

Before the Scientific Revolution, geometry, astronomy, and kindred studies all ranked below theology. (From Margarita Philosophica, *by Georg Reisch, 1508)*

little more than a servant to the study of medicine, engineering, and astrology—became a clearly separate subject in the academic world. By the end of the 16th century, mathematics had begun to be studied to some degree for its own sake. The translation of and interest in Greek mathematics increased with the resurgence of classical literature of all sorts during the Renaissance. The printing press, invented over a century before Copernicus's theory was being circulated, helped establish mathematics as a serious and effective scientific occupation.

The teaching of mathematics became an important aspect of the profession in 15th- and 16th-century Europe. This period, being both the Age of Discovery and the beginning of the commercial and urban revival of the Western world, found great need for the application of measurements and simple calculations for exploratory and trade navigation, as well as for the surveying of land. Computational methods, notably multiplication and division, were first devised for the business needs of the new, rapidly expanding middle class. Along with improved bookkeeping methods, they were seriously studied, especially by those in the banking and finance industries. It was these *bankers* and *financiers* who largely directed and stimulated the great Commercial Revolution, for most people in business, navigation, or sea transport had little knowledge of even the simplest forms of arithmetic.

Universities, private schools, and academies—all of which were still run mostly under church auspices—had little interest in teaching such practical studies. However, with the advent of the Reformation, and the subsequent, new religious justification for improving one's earthly lot through industry and education, schools became more responsive to the realistic needs of the middle class. Many began to hire mathematics teachers, as did scores of the burgher guilds' trade schools in northern Europe. By the time of the French Revolution, most schools—even church-operated and grammar schools, dedicated traditionally to the teaching of only classical and humanistic literature—included mathematics as a major portion of their curricula. Special business, trade, and vocational schools soon became popular; these often specialized in practical mathematics. By the 19th century, *normal schools* had been established throughout the West for the training of public school teachers to meet the needs of the newly established public schools. The teaching of mathematics had become a highly sought-after and fairly secure—although not very lucrative—profession.

For a time, the development of mathematics continued to take place mostly because of the interest in astronomy. The 17th century was a turning point of sorts for the field, in terms of advances in theory, application to the higher sciences, and acceptance by the general scientific and academic communities as a distinct discipline worthy of standing on its own merits. Galileo, Johannes Kepler, and Copernicus all used mathematical computations in their epoch-making calculations of the motions of heavenly bodies. John Napier invented *logarithms*, which revolutionized trigonometric calculation, while Thomas Harriot and William Oughtred developed new algebraic expressions and methods. However, François Viete (Vieta) is considered the "father of modern algebra," since he was the first to use letters to symbolize unknowns and variables. Blaise Pascal and Pierre Fermat founded the theory of *probability*, the study of mathematical predictions that particular events may occur and how frequently. Fermat also established the modern theory of numbers through his study of the properties of whole numbers. *Projective geometry* became a field of some interest, owing largely to the sections of cones worked on by Gerard Desargues. Modern *analytic geometry*—based on a union of algebra and geometry—was the brainchild of René Descartes. The work of Descartes led the way to the development of *calculus*, which was devised independently by Isaac Newton and Gottfried Wilhelm Leibniz. Calculus is essentially the application of algebra

The coat of arms of mathematician Johann Stabius includes some tools of his science in the upper right. (By Albrecht Dürer, c. 1520)

to changing phenomena (things or events), such as the accelerated motion of celestial bodies that so intrigued Newton. Since these motions were represented geometrically by curves, calculus was actually a further refinement of Descartes's analytic geometry.

Modern mathematics is thought to have begun with the work of Newton, who designed a few simple and basic laws of motion that essentially explained in mathematical terms the workings of the universe. Newton's work was a sort of culmination of the Copernican Revolution, since his revelations proved that the Copernican system was sound and accurate. The power and broad applicability of mathematics were clearly seen in Newton's work, for with a few basic axioms, he was able to explain the fundamentals of mechanics, physics, and astronomy. From this point on, there was a growing belief that reasoning capabilities—particularly axiomatic and mathematical deduction—could conquer all the unknown worlds in any sciences, and perhaps even the philosophies. Once mathematics had clearly been extended beyond Euclidean geometry, the way was open for the *Age of Reason*.

From the 18th century on there was a continual refinement of algebra and geometry, while the development of calculus led to the further study of mathematical analysis or *pure mathematics* (a concept that has been commonly used only in the last two centuries to differentiate it from *applied mathematics*). Mathematical analysis was greatly popularized in the middle of the 18th century through the works of the prolific Swiss mathematician Leonhard Euler. The first attempt to systematize this approach was made at the end of that century by French-Italian astronomer Joseph Louis Lagrange, who studied perturbations (gravitational effects of celestial bodies other than the Sun upon one another) through elaborate mathematical analysis. Others, like Pierre Laplace and Baron Fourier, augmented Lagrange's work in the early 19th century. The real foundations of systematic analysis were set a half a century later with the advances made by great mathematicians such as Augustin-Louis Cauchy, Carl Friedrich Gauss, George Riemann, and Karl Weierstrass. These men all applied detailed analysis to prove or disprove theorems concerning physics, mechanics, astronomy, and even numbers themselves. Gauss used analysis to show that a polygon of seven sides (a heptagon) could not be constructed with compass and straightedge, opening the mathematical adventure of deducing proofs of the impossible. Riemann created mathematical theorems based on a non-Euclidean geometry, a notion that had at one time been considered absurd.

Geometry was finally expanded in the 19th century to take into account serious formulations of non-Euclidean geometry. Girolamo Saccheri had come close to establishing such a viewpoint a whole century earlier, and Gauss actually created such a system before the end of the 18th century, but was reluctant to publish it. The credit for this milestone in the history of mathematics goes formally to two independent researchers, Nikolai Ivanovich Lobachevsky of Russia and Johann Bolyai of Hungary. Lobachevsky, the son of a peasant, was so brilliant a mathematician that by the age of 30 he was appointed president of the University of Kazan. He quickly achieved fame as a mathematical revolutionary who insisted that the Euclidean axioms were not always as self-evident as had been assumed for centuries. He took special issue with Euclid's fifth axiom, which essentially stated that "through a given point, not on a given line, one and only one line can be drawn parallel to the given line." Lobachevsky eventually showed that by using a non-Euclidean geometry at least two such lines could be drawn. He began with an arbitrary axiom that ran counter to what was accepted in Euclidean geometry. This axiom stated that "through a given point not on a given line, at least two lines, parallel to the given line, could be drawn." Using this axiom, he was able to devise a mathematical system that had its own inner consistency and logic. Bolyai came to similar conclusions, unaware that such findings had already been published. Riemann and others soon embellished the new idea with their own non-Euclidean geometries.

Modern mathematicians had been freed from the

stranglehold of the Euclidean framework. One result of this new freedom was Albert Einstein's theory that all motion was relative to the analytic framework within which it was reviewed. Einstein's general theory of relativity, stated in the first half of the 20th century, specifically demonstrated the interrelationship of mass and energy, analyzed elaborately, but beginning with the simple algebraic expression that is probably the most famous of all equations: $E = mc^2$, where E is energy, m mass, and c the velocity of light. (Einstein, incidentally, thought Riemann's model most closely approximated reality.)

The greatest significance of the establishment of non-Euclidean geometry was not simply in proposing a better or more workable mathematical model than the Greeks had known. It was that, from this point on, mathematicians were made keenly aware that no mathematical models, regardless of their axiomatic bases, could be accepted as mirror images of reality. They are models

Mathematics has long been a staple in the school curriculum. ("The Arithmetic Lesson," from Frank Leslie's Popular Monthly Magazine*)*

which *may* represent something real in nature, but they were designed only as self-consistent systems, from which workable hypotheses concerning reality could be posed and analyzed in rigorous fashion.

William Rowan Hamilton, an Irish mathematician, soon made the same case for algebra. He discovered that he could devise a logical algebra that, while different from traditional algebra, could be instrumental in helping solve a specific problem that had eluded solution by the long-accepted commutative law of multiplication. This law states that A times B is always equal to B times A. Hamilton began his assertion with the arbitrary, noncommutative axiom that held that B times A could equal A times B. Today, *abstract algebra*, known as *noncommutative algebra*, helps to solve many problems in different sciences, even though it is based on apparently false premises. Again, the significance of mathematics has been shown to lie within the realm of internal consistency rather than absolutism.

Mathematics today is based largely on *formal axiomatics*, which entails the understanding that postulates, laws, and axioms are only relative in value or meaning to the system within which they are created. This is essentially considered the realm of pure mathematics. *Material axiomatics*, a view that assumes that certain axioms or laws are self-evident and eternal, is generally considered to lie within the realm of *applied mathematics*, as when a scientist uses a mathematical model in full confidence of its objective and absolute qualities.

Most mathematicians of the 18th and 19th centuries were considered impractical beings, who had little to offer the real world. Some were patronized by fashionable monarchs, who took pride in the academic reputations of their courts. For a time they had been employed as servants to astronomy, but as that science established greater independence, mathematics had to stand alone. Many a great mathematician received royal and academic applause, and sometimes even compensation. But many others lived in dire poverty or struggled daily with part-time jobs or titular appointments in museums, libraries, and the like, under an unstable patronage system. We know, for instance, that Gauss—one of the greatest mathematicians of his time—found himself without a job when his sponsor, Ferdinand of Brunswick, was killed in a battle against Napoleon. Fortunately Gauss's friends were able to arrange a job for him as the director of the Göttingen Observatory in 1807, but he lived in relative poverty for years. Other were even less fortunate.

Universities have provided the most secure employment situations for mathematicians, but with the advent of the Industrial Revolution mathematicians found themselves in demand for private research and industry as well. Today, while many mathematicians continue to work for universities, they can also look to governments, private industries, and businesses for substantial employment opportunities. With the tremendous boom in science and technology in the 20th century, the professional and social status of mathematicians has soared, as have their earning power and job security. Far from being considered impractical, mathematicians are now very highly regarded in the academic, industrial and commercial worlds. They are widely sought and very highly paid in many areas, especially in fields related to nuclear physics and other modern sciences. They are employed as consultants to business and government, particularly in economics and technology. They are important figures in all of the electronic and computer industries, and figure prominently in the vast space and underwater explorations of the greatest nations in the world.

There are two main approaches to investigation in mathematics today. *Applied* or *practical mathematics* is concerned with analyzing or quantifying problems in the real world by making analogies to it with mathematical objects, which are more readily worked with. *Pure mathematics* is the study of mathematics as a system of logic capable of illuminating the nature of absolutes. In modern times, "absolute" mathematical laws are often applied to seemingly paradoxical situations in less objective sciences, in the hope that the apparent paradox may be resolved by the absolute laws of mathematics. The result is the recent trend of mathematical, quantificational, and statistical analysis applied directly to what were traditionally viewed as speculative sciences. Since the 19th century, when mathematical models first became widely used to complement (and too often replace) philosophical speculation, intuition, and insight, the process has been thought of as the maturing of science in general. This process has been rapidly and suddenly enhanced in the 20th century with the development and increasingly broad use of computers and other sophisticated technology. But mathematicians did not separate themselves into *pure* and *practical* camps until the second half of the 19th century. Before that, the same mathematician who explored the nature of mathematical objects, such as numbers, and fundamental laws and principles, also made direct application of these things to problems in logic as well as to such worldly areas as technology and engineering.

Today we find that very few practical people who apply mathematical principles and models to their special investigations—a particular chemical analysis, for example, or a genetic codification scheme—are really mathematicians. That is to say, they know little or nothing of the derivation of mathematical symbols, principles, or laws; they only assume these to have been perfectly worked out and ready for application. In one sense, the only real mathematicians in the contemporary world are those concerned with mathematics as its own unique scientific system. Mathematics is not pure or absolute, however. The symbols used in the science may be internally coherent, given a particular model, and the axioms or principles derived from them may therefore be logical and even necessary. But the model itself must always be questioned, as mathematicians themselves know all too well, if one seeks to propose it as an absolute picture of reality. Also, any mathematical model, principle, or law refers only to a given and contrived situation; it is not necessary that this situation mirror the natural world, only that it be consistent within itself. Mathematics, then, may greatly enhance the objective and quantifiable scope of any science, but it cannot ensure the derivation of absolute quantities or put an end to speculation and interpretations—though even scientists often think it can.

Although applied mathematics is the most visible aspect of the profession today, great strides have also been made in the realm of pure mathematics. Since World War I, an influx of ideas has so reshaped our very conception of mathematics that it is now common to speak of the major perspectives as "philosophies of mathematics"—an interesting term, since philosophy and mathematics are generally regarded as opposite sides of a coin. These philosophies have been propagated by some of the great thinkers of our century: Jules Henri Poincaré, George Cantor, David Hilbert, Alfred North Whitehead, Bertrand Russell, and Einstein. They are generally significant in that they have clearly defined mathematics as a relative study and tool, using logic and operations that are totally symbolic and convenient, but rarely natural and never absolute. Hilbert's formalist thesis, for instance, underlined the symbolic and relative nature of basic mathematical terms from which extensive axiomatic systems and operations are constructed. Russell and Whitehead championed the logistic thesis, which emphasizes the necessity of internal consistency rather than the natural or absolute value of the same terms. Einstein made use of non-Euclidean geometric forms in arriving at his unique concepts in physics.

All of these philosophies made a sharp break with the tradition of the earlier mathematical theories, which held that the field was the purest and most absolute discipline in science. Suddenly mathematics came to be seen as only a base from which different views of reality may—more or less tentatively and cautiously—be proposed. In this vein, mathematics is thought of as a tool and an operation, much more than as some sort of a microcosm of universal law. In fact, the variety of symbols and meanings that contemporary mathematics is endowed with is so vast and elusive that Bertrand Russell was once prompted to conclude that "mathematics may be defined as the subject in which we never know what we are talking about, nor whether what we are saying is true." While this may appear to be a degradation of the science from the absolutist standpoint, it can also be seen as a tremendous advance looked at from the perspective, as Georg Cantor once remarked, that "the essence of mathematics lies in its freedom"—that is, its ability to be manipulated, to be the most useful in the specific contexts in which it is most needed at any given time. Its truest value seems, then, to be rooted much more in its fluidity than in the absolute rigidity that it was once assumed to reflect.

For related occupations in this volume, *Scientists and Technologists*, see the following:

- Alchemists
- Astronomers
- Astrologers
- Biologists
- Chemists
- Computer Scientists
- Engineers
- Physicists
- Statisticians

For related occupations in other volumes of the series, see the following:

in *Communicators*:

- Scribes

in *Financiers and Traders*:

- Accountants and Bookkeepers
- Bankers and Financiers

in *Healers* (forthcoming):
  Physicians and Surgeons
in *Leaders and Lawyers*:
  Political Leaders
in *Scholars and Priests* (forthcoming):
  Priests
  Scholars
  Teachers

Physicists

It was assumed for many centuries that the study of physics applied broadly to the investigation of the natural world in all its complexity. By this definition most *scientists* were *physicists*. But since the Scientific Revolution, which took a firm hold on Western thinking by the beginning of the 17th century A.D., physics, like all other sciences, has come to represent a more particular field of study. Specifically, physics is now spoken of most commonly as the science of dynamics or motion. In answering the question "How do things move?" physicists have also found themselves (often unwittingly and even reluctantly) at the center of a philosophical and theological controversy.

More comfortably, and in keeping with their avowed interest in the physical world itself, physicists have most recently been likely to investigate the nature and composition of matter and energy. The characteristics of matter are key ingredients in the study of motion, because it is matter that is altered by dynamic events. The nature of energy must also be discussed in relation to its absolutely essential role in any concept involving movement and action.

It is easy to see that the concerns of the physicist overlap considerably with those of other scientists, especially the *chemist, biologist,* and *astronomer.* Physics is also of ultimate concern to the *mathematician*, since it, probably more than any other science, takes an essentially mathematical approach. Findings in the field are almost always expressed mathematically, because they are far too complex for ordinary language. Furthermore, even these mathematical axioms are constantly being summarized and generalized according to simple principles of law (as few as possible) of the physical universe.

The ancient Greeks gave physics its first real scientific foothold with their mathematical developments, notably in their establishment of axioms and mathematical proofs. These made mathematics a valuable tool in testing hypotheses and deriving general principles. Physicists put this tool to use, particularly in astronomy, a field that preoccupied most physicists until modern times. Since the heavenly bodies could not be directly observed by the Greeks, they had to be studied through analogy to mathematical models or else through sheer speculation.

Although the study of the heavens was a serious business in the ancient world, which thought planets and stars to be divine essences, planetary motion as such was little cared about. Most astronomers and *philosophers* thought that the Earth and stars were stationary; they explained the movement of the Moon or the other planets in terms of divine whim. The Pythagoreans and Aristarchus were the first to propose the concept of the motion of the Earth, and they made the earliest attempts to understand the phenomenon. Yet they did not base their studies on very sophisticated notions of physics, but rather more on vague speculations in astronomy.

Archimedes, one of the greatest of the ancient mathematicians, studied buoyancy, levers, and balancing points, but more important, he established mathematical methodology as the best technique for analyzing quantities and distances in the physical world. But the greatest strides in ancient physics were taken by Democritus, the Classical Age philosopher from Thrace. He is most famous for this atomic theory, which held that all matter is made up of basic and changeless particles called atoms. What distinguishes one form of matter from another, according to this theory, is simply the arrangement of the atoms. However, few people at the time even thought of Democritus's atomic theory. Not until modern times would scientists understand that, in its fundamentals, the theory was largely correct.

During the Middle Ages there was little progress in physics except in a few categories of mechanical theory, such as the investigation of balances and levers. As in ancient times, there were still no real physicists, only astronomers, *physicians*, and *philosophers* who had passing interests in the subjects of motion and material structure, as related to their primary research. Although not

Archimedes is said to have discovered the principle of the displacement of water while in his bath. (From Gaultherius Rivins, Architecktur . . . Mathematischen . . . Kunst, *1547)*

yet an independent field, physics was an important part of science nonetheless, especially astronomy. Probably the greatest advance in this period was made in the investigation of projectile motion—what is now referred to as *impetus theory.* Essentially, scientists wanted to know what sustained the movement of an object once it had been projected through space. For instance, if a spear was thrown toward a target, it was quite clear that the original impetus came from the thrower. The problem was in figuring out why the spear was able to continue to the target after the hand had released it. Centuries earlier Aristotle had posed a solution to the problem; he thought that a projected object split the air, which then curved around behind it and gave it a continued push for some distance.

Aristotle's explanation of projectile motion was not seriously questioned until the 14th century A.D. by a group of European scholars led by Jean Buridan. Buridan first demonstrated that a grindstone, having been started in motion, continued to turn for some time after the initial push had ceased. This was significant because no currents of air could be detected around the rotating stone. Even more conclusive was the fact that there was no backside to a round grindstone, so that there was nothing against which circulating air currents could push. In a similar experiment, Buridan showed that a spear sharpened at both ends traveled the same as a regular spear, even though it had no blunt side against which air could push. Buridan concluded that a projectile had to be set in motion by an external force, but after that, its continued motion was not being caused by anything external. It was, rather, driven by a motive power that was inherent in the object itself, a power that Buridan called *impetus*. He even determined that the impetus of a projectile is proportional to both its velocity and weight.

The research of Buridan and his associates represented the first authentically experimental work done in the field of physics. It was done in the style of Roger Bacon who, a century earlier, had insisted that scientific investigation must be experimentally verified or rejected, not merely judged according to argument, persuasion, or theological dogma. This new scientific method was important for the establishment of all the modern sciences, but especially physics, since physics alone rested totally on mathematical, and therefore purely logical, foundations.

The scientific method created tension with the church, however. At that time the church still largely controlled European education and scholarly research, from the lowest levels right up to the university, which was just beginning to exert a force on Western intellectual development. Although much of Bacon's work had been done at the request of Pope Clement IV, most of his writings were eventually banned by the church, which was shocked to realize that he was proclaiming human reason as capable of uncovering mysteries in the universe. The church insisted that it alone had access to such things and that its dogma was based squarely on divine revelation. Reason and intellect could not compete with such authority, the church said, and it was blasphemous for them to try.

Physics in particular was clouded by church investigations, such as that which proclaimed that the planets were kept in perpetual motion by attending teams of angels. Buridan's pioneer research in projectile motion clearly refuted such a notion, offering instead a scientific and experimentally verified explanation of the perpetual orbiting of planets. But Buridan, whose work bore such insight that it anticipated ideas that would be fully expressed by Newton's first law of motion some three centuries later, was careful to try to blend his findings quietly into the broadest Christian worldview. In relating his new evidence for projectile momentum, to celestial motion, he put his scientific research in a religious context:

> When He created the world [God] moved each of the celestial orbs as He pleased, and in moving them He

expressed in them impetuses which moved them without His having to move them any more . . . . And these bodies were not decreased nor corrupted afterwards . . . Nor was there resistance which would be corruptive or repressive of that impetus. But this I do not say assertively but so that I might seek from the theological masters what they might teach me in these matters as to how these things take place . . .

During the Renaissance, physicists became bolder in asserting mechanistic theories to explain both earthly and heavenly motion. The Scientific Revolution had been sparked by the assertion of Nicolaus Copernicus that a rotating Earth was not the center of the universe, as had long been assumed. This struck down the church's teaching that mankind stood still at the perfect center of the universe, which moved about it. Moreover, this startling discovery had been made by way of scientific reasoning, not divine revelation. The die had been cast, and future scientists were to base their scientific findings on intellect and reason, much to the dismay of the church and its theologians.

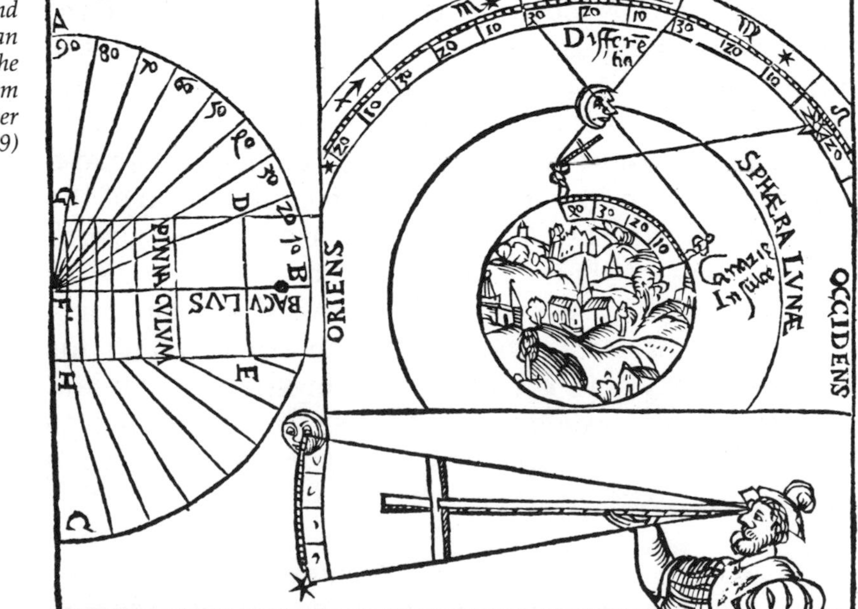

The work of astronomers and cosmographers had an enormous influence on the history of physics. (From Cosmographia, *by Peter Apian, 1539)*

Great strides were taken in physics in the wake of Copernicus's new methods. Motion studies were advanced chiefly through careful observations of free-falling bodies. These studies were greatly enhanced by the application of mathematical principles, a new method of analyzing scientific data. Galileo may be most famous for his astronomical observations, but his greatest contribution to physics was his systematic and quantitative study of the nature of the motion of free-falling bodies. He discovered that bodies fall with constant acceleration, and so proposed the *law of inertia*: In the absence of interference, like friction, an object on a horizontal plane continues moving without altering either speed or direction.

Galileo was the first abstract theoretician in physics. After his work physicists treated this subject quite seriously, realizing that in it might lie the answers to the workings of both Earth and the heavens. The value of Galileo's studies in the field was that he used simple experiments on everyday objects to derive abstract principles. For instance, by rolling balls down inclined planes of various slopes, he was able to see that their motion accelerated at a uniform rate. From this observation he was able to elaborate mathematically that velocity and distance increase uniformly with time at a fixed ratio. Later, these ideas were to form the basis of Isaac

The 16th century saw numerous experiments on laws of motions—here, on the path of a cannonball. (From Problematum Astronomicorum, *by Daniel Satbech, 1561)*

Newton's monumental work in physics, astronomy, and mathematics.

Newton's three laws of motion were grand elaborations of Galileo's simple experiments. Newton placed greater emphasis on the role of force in motion than anyone else ever had. Force, according to Newton, was relative not only to direct contact, but also to weight, resistance, and gravitation. On a larger scale, Newton applied his principles of physical motion to his view of the universe. Accordingly, he calculated that universal gravitation varies inversely, as the square of the distance between two bodies, and directly, as the product of the masses of the two bodies. This law laid the foundation for Newton's mechanistic image of the heavens. Physics, then, had lent keen insight into the age-old issues concerning the workings of the universe. It was largely through the works of Newton that physicists came to understand the world in mechanistic terms, whereby the attraction, mass, and dynamic velocities of all planets and heavenly bodies affect the motions and positions of one another. Understanding the complexity of the universe and the physical laws governing it, physicists thereafter sought to refine, modify, and elaborate Newtonian physics in light of the new age of astronomy.

While Newton was so greatly affecting the future progress of astronomy, others were investigating physics from a more practical stance. Some were working out the principles of the vacuum, for instance; these studies eventually led to the building of the steam engine, which was destined to become the backbone of the Industrial Revolution. These physicists, then, had a great influence on the course of history and the shaping of society. Not surprisingly, they were for the most part recruited from among the privileged and aristocratic classes. While they openly supported the status quo, their collective pursuit in physics went as far as that of any other science or political philosophy in overturning the old social and political order. Their findings led directly to the technological innovations that made the Industrial

Galileo paid homage to his predecessors, Aristotle, Ptolemy, and Copernicus, in his great work. (From Dialogue on the Two Chief World Systems. . . , *1642, Leiden)*

Revolution possible and practical, so they had a prominent role in the establishment of mass production, capitalist economies, and democratic societies.

The physicists of the 17th and 18th centuries helped overturn the old order by indirectly weakening the idea that kings had a divine right to rule nations at whim, and the notion that people were born into certain positions in society and ought not to depart from them, lest they corrupt the entire society and the "great chain of being." These fundamental religious views were undermined, usually unwittingly, by physicists' discoveries indicating that the universe was held together by physical laws of motion and atomic structure—the so-called *mechanistic* view of the universe. The older, *vitalistic* view held that the universe and the living beings in it were not simply "machines" but were motivated by some essential living spirit, often thought to be divine in nature.

The triumph of the mechanistic world view over the vitalistic one was far from the object of the research of the early modern physicists. In fact, most early physicists—including Newton himself—were also ardent supporters of orthodox Christianity and would never have wished anything to interfere with it. But their findings were strictly physical ones, which seldom made any direct reference to politics or religion. And the long-range progress of physics—independent of the individuals who forged it—eventually led to a more materialistic and

In this famous physics experiment, two teams of horses were unable to pull apart two hemispheres with a vacuum inside. (From New Experiments, *by Otto von Guericke, 1672)*

democratic view of the state and society. This view was to carry the Western world into a new industrial, technological, and individualistic age, which may be said to have begun in the 19th century and which continues to evolve today.

Physicists even before Newton's time had begun to experiment with physical principles that would pave the way for the industrial and technological revolutions of the future. Otto von Guericke, who was 40 years old when Newton was born, attacked the age-old assumption derived from Aristotle that "Nature abhors a vacuum." While acting as burgomaster of Erfurt in Germany, he delighted himself and crowds of spectators with a series of famous demonstrations. They were designed to show that air, which he had experimentally determined to have measurable weight and density, could be pumped out of a container to create a vacuum. His lavish demonstrations cost him some $20,000 to stage, but he also put them on an exhibition tour from which he probably earned a tidy profit. His most famous experiment, with the "Magdeburg Hemispheres," was demonstrated for the Holy Roman emperor, Ferdinand III. It featured two bronze spheres from which all the air had been pumped out and which could not, as Guericke had promised, be separated by the two eight-horse teams that tried to pull them apart. At other times this scientist with a flair for showmanship had shown that, in a vacuum container, a candle could not burn, an animal could not live, and a bell could not be heard.

Much of Guericke's work had been built upon the research of a far quieter man, Evangelista Torricelli, who had published earlier reports likewise indicating that air had weight. Robert Boyle was doing similar work in England, where he was a member of a scientific society called the Invisible College. That society soon became known as the Royal Society and bore the motto "Nullius in verba," or "Nothing by mere authority." The scientific societies that thereafter attracted the membership of some of the great pioneer physicists of the 17th and 18th centuries pointed to this motto in distinguishing themselves from the church-operated universities, which were bound very much to religious doctrine at all costs. Boyle was considered a rather strange physicist for his time because he insisted on deriving his claims strictly through experimental methods. Abhorrent as this idea was to church authorities, Boyle was always devoutly religious, as were a great many other physicists.

The broad significance of the work of these and other physicists of the time was twofold. First, their discoveries led to many technological innovations and theoretical advances that were to take the Western world into the Industrial Revolution and beyond. The invention of the steam engine, for example, would not have been possible without a knowledge of vacuums. The second impact of these experiments—more subtle perhaps, but fundamentally greater—was that they posed questions (and gave provisional answers) concerning the nature of air, space, and nothingness. What came out of the fact that air was *something* was a redirecting of the efforts of physicists, chemists, and biologists alike. Until then, scientists had assumed that light, electricity, sound, and magnetism were beyond comprehension. Of course they had long been curious about these phenomena. Even in the 16th century William Gilbert, the court physician to Elizabeth I of England, declared that the Earth itself was a great spherical magnet. In the next century the Jesuit physicist Francesco Grimaldi showed that light could actually bend around an obstacle, a phenomenon he called *diffraction.* Guericke proved that electricity actually traveled, and Benjamin Franklin even determined that it had both positive and negative attractions. Experiments like these laid the basis for the modern understanding of the strangely behaved phenomena. Not until the 19th century, however, were physicists—building on the earlier research concerning the nature of air itself—able to relate the phenomena to wave and energy theories, rather than particle theories.

The period from the publication of the Copernican

theory in 1543 until the late 1700s was one in which physicists were preoccupied with the principles of motion and various particle theories, which attempted to explain the structure and composition of matter. By the end of that period and into the 19th century, physicists became increasingly interested in applying these theories and principles to practical concerns. There were three outstanding reasons for this. First, the Industrial Revolution was under way and the great Western economies—particularly those of England first, and North America later—came to depend on mechanical innovations that would support the new means of production.

A second reason for the new focus of physicists was that the new manufacturing conglomerates began to independently sponsor in-house physics research that would directly benefit their modes and rates of production. These capitalists did not order the physicists they employed to do research that might interest the government or military institutions, nor did they ask them to elaborate on astronomical insights or develop new theories of the universe. Their goals were specifically and totally utilitarian. Given this situation, it is clear that the traditional stereotype of the aristocratic gentleman-scientist who dabbled in physics as much as chemistry, biology, astronomy, and probably even some of the liberal arts, was destined to become quickly outdated. Times were changing, and in capitalistic industrial nations there was suddenly a great emphasis on immediacy, innovation, and practicality.

The third reason for the changing focus in the field is perhaps the one that most concretely transformed physics from a genteel pastime or scholarly amble into a rigorous profession, grasping the challenges of the new age head-on. This was the new recognition within colleges and universities of physics as a field of research somewhat independent of chemistry, biology, and astronomy. What this came to mean, of course, was that students could be trained specifically as physicists. Moreover, the 19th century was one of great and broad educational reform, which opened wonderful new opportunities for many more people from the middle classes. It was no longer necessary to be a person of independent means or political connections to obtain an education and eventually a job in physics or any other field. Toward the end of the 19th century and into the 20th, this notion of universal education was gradually extended to all social and economic groups, including women and minorities, in the industrialized and developed nations of the world. Physics for the first time became a significant profession, as the democratization of education required physicists in the classroom at secondary and college levels.

Perhaps the greatest work of physicists from the 19th century to the present day can be very simply capsulized by looking at the progress of research in electricity, in magnetism, and in the role played by energy in thermodynamics. Advances in academic or pure physics have often come in the wake of interpreting some of the applied research in the field.

The quantitative study of heat and its mechanical action—now referred to as *thermodynamics*—really dates back to Galileo and his early invention of a crude thermometer. This instrument allowed Galileo to speak of heat not only qualitatively, but also in terms of relative *degrees*. Yet it was almost two centuries before significant advances were made in the quantitative study of heat. A report published by the Scottish physicist Joseph Black in 1803 pronounced on the principle of thermal equilibrium:

> . . . we can perceive a tendency of heat to diffuse itself from any hotter body to the cooler ones around it, until the heat is distributed among them in such a manner that none of them is disposed to take any more from the rest. The heat is thus brought into a state of equilibrium.
>
> This equilibrium is somewhat curious. We find that, when all mutual action is ended, a thermometer applied to any one of the bodies undergoes the same degree of expansion. Therefore the temperature of them all is the same. No previous

> acquaintance with the peculiar relation of each body to heat could have assured us of this, and we owe the discovery entirely to the thermometer. We must therefore adopt, as one of the most general laws of heat, the principle that all bodies communicating freely with one another, and exposed to no inequality of external action, acquire the same temperature, as indicated by a thermometer.

Black went even further in discovering that different substances have different "heat capacities"; that is, that equal masses of different substances will undergo a different change of temperature upon acquiring or losing the same quantity of heat. He was then able to decipher the heat capacity per unit mass, or, in other words, the *specific heat* of a substance.

Until the middle of the 19th century physicists generally believed that units of heat—*calories*—were actually a substance of some sort, most likely a fluid. There were those who objected to this caloric theory, preferring instead the notion that heat was nothing more than the rapid movement and loosening (expanding) of the particles that made up all substances. Early in the century Count Rumford showed that the mechanical rubbing of ice caused it to melt, from which he deduced that mechanical motion was actually converted in heat. Furthermore, since the ice and melted water had the same weight, heat apparently had no substance of its own. Scientists as far back as Newton and Descartes had suspected as much, but it was left to those 19th-century physicists concerned with the mechanical output of machines and engines to finally put the caloric theory to rest.

James Joule, son of a middle-class brewer, discovered that the amount of work that entered a system was closely related to the amount of heat that came out. Sadi Carnot had done similar work earlier while trying to reduce the heat loss, and thus the inefficiency, of engine systems. He used the term *motive power* to indicate the amount of work that could be obtained from a given amount of heat. This heat he determined to be nothing other than *friction*, the vibrational motion of particles. Having been the first to quantitatively consider the way in which heat and work are freely exchanged, he is considered the father of thermodynamics. In determining that motive power could neither be created nor destroyed in nature, he fully understood and demonstrated the first law of thermodynamics, containing the principle of the *conservation of energy.*

Thomas Young had introduced the use of the word *energy* in relation to thermodynamics in 1807, and further experiments concerning the nature of energy by James Clerk Maxwell finally destroyed any remaining credibility attached to the caloric theory. Heat was clearly a form of energy. Joule studied it in various forms: electrical, mechanical, and thermodynamic. Rudolf Clausius studied the process of converting heat into work, a process he believed would eventually reach a perfect equilibrium on a universal scale. This would mean that all temperatures would become equal and no energy could exist, a pessimistic concept referred to as the "heat-death of the universe." Maxwell along with August Karl Kronig studied energy even further in developing the kinetic molecular theory, which describes the random action of large numbers of particles. This led to the establishment of *statistical mechanics*, allowing quantitative predictions of energy processes and their outcomes. Based largely on these studies, the American researcher Josiah Willard Gibbs developed the science of chemical thermodynamics, which today forms the basis of the modern chemical industry.

Along with the quantitative studies of heat, thermodynamics, and energy, physicists of the 19th and 20th centuries found themselves engrossed in elaborate research on the nature of electricity, magnetism, and light. The work in these fields has been immense and complex but can briefly be summarized as follows:

When physicists first began to see light as an energy phenomenon that traveled in waves, rather than as particles or projectiles moving directly from a source to a receiver, they tried to determine through what medium it was transmitted. In the 17th century Christian Huygens had called the medium *aether*. The term was borrowed from Aristotle, who declared that this fifth element was the element of the heavens. Physicists had considerable difficulty in assigning aether enough substance to transmit light, yet not so much that it would distract or perturb planetary motions, which were then being closely scrutinized by astronomers and physicists alike. However, Huygens's work did later inspire that of Augustin Fresnel, Thomas Young, and others, who largely established the wave theory of light by the middle of the 19th century.

A pioneer in the fields of magnetism and electricity was Michael Faraday, a prototype of the new physicist. Born into a poor laboring-class family in 1791, he could not even afford a decent education. Instead he became an apprentice in a bookbindery, where he educated himself. He eventually received an appointment to a professorship at the Royal Institution. In that capacity he soon earned a reputation as one of the greatest physicists and chemists of his time. Among his many accomplishments in science, he learned that there were distinct lines of magnetic force, which were similar to electric lines of force. He even supposed that it was along these lines that light traveled, rather than through some mysterious aether. He showed conclusively that electricity was not a fluid and did not operate along simple mechanical lines.

Later in the 19th century, James Clerk Maxwell was conducting experiments that showed that magnetism and electricity were interdependent. He, too, supposed that electromagnetism was made up of lines of force rather than any aethereal substance. The *electromagnetic theory* was a revolutionary concept in proposing that perhaps light, electricity, and heat were energy forms which had no material existence at all—at least not in the way that scientists were accustomed to. In suggesting that there was an oscillation of an electric charge, which created its own emanating electromagnetic field, Maxwell established the *field theory of energy*, which was eventually accepted.

It soon became clear that, in their search for practical knowledge, modern physicists were uncovering some of the great secrets of the physical universe, leading to fundamental reworkings of basic theory in the realm of both pure and applied physics. Heinrich Hertz did work on radio waves at the close of the 19th century, which verified the wave and field theories of Faraday and Maxwell. New electromagnetic fields were then discovered by Wilhelm Roentgen, who found the X-ray region, and Henri Becquerel, who then found the gamma ray region, elucidating the full electromagnetic spectrum. Ernest Rutherford determined that radioactive rays, unlike the X-rays discovered by Roentgen, were complex. He and Joseph John Thomson realized that radiation was actually like a microcosm of the solar system, with *beta rays* eventually being found to be electrons and *alpha rays* found to be a nucleus. These discoveries were fundamental to the quantitative studies of atomic structure undertaken by Niels Bohr in the mid-20th century.

The 20th century has been the period of greatest advancement in theoretical physics. Much of the progress has been aided considerably by new instruments of observation. The *spectroscope* is perhaps the greatest of these. It differentiates various light wave frequencies in such a way that it can actually determine the inner workings of the atom.

In the early years of the 20th century, Max Planck worked out a theory of energy that would lead to a rethinking of traditional physics and would even modify the newly accepted field theory of energy. His work was basic to the further elaborations of Albert Einstein, whose special and general theories of relativity eventually challenged even traditional concepts of time and space, employing whole new mathematical systems. Planck's simple observation was that energy is not infinitely subdivisible, but, like matter, had some particle nature. The particles of energy, to be distinguished from atoms, he called *quanta*. The *quantum theory* represented a virtual combining of particle theory and field theory. It proposed a matter-field dualism that seemed to give some credence to both Newton, who had thought of energy in terms of particles, and Maxwell, who determined that it existed expressly in emanating fields. Quanta were thought of as impact particles moving at incredibly high speeds. Hertz had indicated that light, for instance, possessed both wave and particle natures. This was shown in his experiments with the so-called *photoelectric effect*, whereby light makes metal electrically active by bombarding it with *photons* (impact particles of light).

Einstein astonished the world of science when he determined that matter was nothing more than a form of energy. In his special theory of relativity he worked out an interrelationship between mass and energy in that famous equation, $E = mc^2$, where E is energy, m is mass, and c is the velocity of light. Mass and energy were thereby interpreted as different aspects of the same phenomenon. This idea was hard to accept in the light of classical physics, even though the wave-particle dualism of photons had already been observed by Hertz and identified by Arthur Compton.

If light waves in fact had high-velocity impact particles, that did not necessarily mean, as Einstein was seemingly suggesting, that the inverse was true—that is, that particles of matter demonstrated characteristics of waves. But in 1924 Louis de Broglie established the existence of such a symmetrical inverse of the *Compton effect* (wave-particle dualism), when he showed that there was also a particle-wave dualism. Working within the framework of quantum physics, he adapted Einstein's formula (which related mass and energy) along with Planck's (which related frequency and energy) to his own conclusion that each particle of matter has its own associated wave, today called a *matter wave*. Other physicists

Men and women physicists—like Lise Meitner and Otto Hahn, here—worked together as equals in the 20th century. (American Institute of Physics, Niels Bohr Library)

in the 1920's actually detected various wavelengths associated with particles. It was established then that matter and energy are interchangeable: Particles are always wave-like, and waves are always particle-like. These findings finally made sense of Einstein's research, which literally restructured classical physics.

In the period between the two world wars, quantum theory may well have made the profession of physicist the most exciting one in the history of science. Physicists of this era—both men and women—are best remembered for their pioneer work in atomic energy and for the construction of the atomic bomb. Most recently, physicists have become closely associated with their work in nuclear energy. They have been called on to develop the most deadly weapons of war ever known to mankind. They have also been called on to develop nuclear energy as a source of productive power for industrial societies at a time when oil and gas have begun to become scarce and unduly expensive.

But these challenges to physicists have not been without cost. Tremendous moral questions have come to bear on the profession, as scientists wonder how far should they go in supporting the military might of their respective nations, and how great a public health risk can they concede and continue to develop nuclear energy for domestic peacetime use. These are questions that physicists over the last several decades have had to answer on an individual basis. Some have even left the field, moving to other areas of science. As a group, though, they have seemingly decided to do whatever they can to make nuclear weapons and energy available, hoping that the military, industrial, and political authorities who obtain them will put them to wise and humane use. This notion has been widely criticized by those who think that the scientific advances made by physicists are in the process of destroying the quality of life on Earth, and perhaps one day may even lead to the destruction of life itself.

For related occupations in this volume, *Scientists and Technologists*, see the following:
Astronomers
Biologists
Chemists
Computer Scientists
Engineers
Mathematicians
Scientific Instrument Makers
Statisticians

For related occupations in other volumes of the series, see the following:
in *Builders*:
Architects and Contractors
in *Healers* (forthcoming):
Physicians and Surgeons

in *Scholars and Priests* (forthcoming):
Scholars
Teachers
in *Warriors and Adventurers*:
Soldiers

Scientific Instrument Makers

The profession of *scientific instrument maker* did not really exist before the 17th century. Before then, science was largely a matter of philosophical speculation and simple, naked-eye observation. There was little concern for any scientific methodology in the sense of proving or even probing the truth about the natural physical world, so there was no need for related theories to be based on quantification studies, measurement, precise calculation, or physical analysis. But as the Scientific Revolution unfolded, professional *artisans* and *scientists* began to labor over the conception, construction, and refinement of scientific instruments. By the 18th century the profession had become truly fine and respectable, as it became somewhat of a fashion among aristocrats and the upper middle classes to own their own scientific instruments with which to pursue their amateur interests in astronomy or geography—or simply to use them as conversation pieces, decorative items, or status symbols, as was much more commonly the case. Most dignified 18th-century noble or merchant families were able to proudly display at least a few such fine items.

Almost all scientific instruments can be broadly classified as belonging to one of two types. One type contributes to observation; these generally include in their name *-scope*, as in *telescope* or *microscope*. The other contributes to measuring, and generally includes the word *-meter*, as in *barometer* and *thermometer*. In ancient times, several types of observational and measuring instruments were in use. Lenses were used rather widely for magnifying glasses. The Greeks and Romans called them "burning glasses," because of their ability to magnify heat as well as light. Other instruments were used, too. Sundials were used in many societies to calculate time during sunlight hours. The Egyptians used a water clock called the *clepsydra* to measure time at night; the Chinese had even more complex water clocks. The Greeks used *astrolabes* to measure the altitudes and determine the positions of heavenly bodies. But these instruments were not widespread enough to warrant the development of specialized makers and, after the fall of Rome, complex instruments were rare in Europe.

The lens was first used for the manufacture of an optical instrument in the 13th century. Shortly thereafter, *opticians* (makers of optical instruments) began to construct crude eyeglasses, which essentially consisted of convex magnifying lenses. These early opticians had little more knowledge about light or human vision than the Greeks had accumulated. The Greeks thought that vision resulted when light—envisioned as a stream of particles called *corpuscles*—transferred the image of an object to the human eye. Whether the light came originally from the object or the eye was a matter of controversy for many centuries. As early as 300 B.C. Euclid had correctly stated the general law of reflection, but the exact nature of refraction remained a mystery until the 17th-century work of Willebrord Snell.

In the 17th century, opticians set the stage for a revolution in the making of scientific instruments. Most of these devices were made to prove scientific theories. As a result, they were often called *philosophical instruments*, for science was then thought of as much more a matter for philosophical speculation than observation, measurement, and analysis of concrete phenomena. But advances in optics soon changed all that. Snell formulated the law of refraction in 1621. Johannes Kepler in the same century described the eye as a lens. Christian Huygens, in his famous 1690 *Treatise on Light*, considered light as consisting of waves rather than corpuscles. The latter view was elegantly defended by none other than Isaac Newton, who learned how to break white light into colors through the use of prisms. In 1704 Newton wrote a very influential treatise on the subject called, simply, *Optiks*. All this "philosophical" speculation led to increased interest in optical instruments. Crude microscopes and telescopes had been developed between 1590 and 1610. Throughout the 17th and 18th centuries, these and other instruments were further refined and elaborated.

At first, scientists themselves made most of their own

Opticians made the precision lenses required by so many scientific instruments. (From Diderot's Encyclopedia*, late 18th century)*

instruments. René Descartes, Huygens, Evangelista Torricelli, Robert Hooke, Galileo, Robert Boyle, and Newton all devised, constructed, and even sold their own instruments. At this early stage they were not willing to entrust such fine precision work and careful lens polishing to artisans and opticians. But scientists quickly gained confidence in artisans, to whom they turned for expertise in the technological aspects of the craft. Together, the two groups set out to make instruments to put scientific theories to the test. The *telescope* would finally show whether or not there was any substance to the Copernican theory of the universe. The *microscope* would once and for all indicate if there really were cellular structures in human and plant tissue. The *air pump* and *barometer* would make it possible to create and measure relatively high vacuums, so that scientists could find out if it was really true that "nature abhors a vacuum." In his lectures in 1603, Galileo used the first *thermoscope* to "show" his students heat and cold as water rose in a special tube. As measurements of this phenomenon were added to the instrument, it became known as a *thermometer*.

Many workshops sprang up in Europe to handle the new demand for scientific instruments. Even amateurs and status-conscious aristocrats purchased such items, often for the sheer joy of collecting and displaying them. Many artisans catered to the more vainglorious group of consumers in their hasty construction of gaudy sundials, decorative astrolabes, compasses, and even *astronomical compendiums*, pocket timepieces that became fashionable for a time. By the 18th century this market had grown to extraordinary proportions. Every worthy prince had his own *cabinet de mathématiques*.

The finest workshops were those that manufactured optical instruments. Holland was the center of this activity for a time, though some opticians working in Italy also earned high reputations. Henry IV retained many such artisans at court, a tradition of patronage that continued for some time in France. But it was in England that the industry came to realize its fullest potential. In London, there were shops specializing in the manufacture of mathematical instruments made of wood and brass; these shops also often made navigational instruments, such as quadrants and nocturnals. As Robert Hooke's works with the microscope became more famous in the second half of the 17th century, opticians rose to the fore in the profession of scientific instrument making. It took hours of painstaking care to properly polish (usually with pumice) the curves on the lens glass to make it a high-quality product. The Dutch scientist Anton von Leeuwenhoek was famous for his expertise as a *lens grinder*, but English artisans soon developed a secret for polishing specula (mirrors or metal reflectors), using an application of amalgams and alloys. This gave the British a firm advantage in the market for optical devices. After 1755 they developed the use of flint glass, which could be used in the manufacture of achromatic (colorless) lenses—a product far superior to anything known at that time. The British also possessed many high-quality raw metals difficult to obtain on the Continent: brass, copper, steel, and tin. Finally, English shipping made sure that these fine manufactured lenses reached even the most distant markets.

The French frequently tried to induce English artisans to set up shop across the English Channel. One instrument maker, Henry Sully, actually built a thriving watchmaking business at Versailles, which employed chiefly British recruits. Perhaps the most significant factor in the success of the British in this profession lies in the fact that their guilds—the Clockmaker's Company for makers of mechanical instruments, and the Spectaclemaker's Company for makers of optical instruments—were relatively unrestrictive. In contrast, such guilds elsewhere were very strict and hindered the growth of the profession. As early as 1565, for instance, before the business of instrument making had even blossomed to its fullest extent, the French monarch Charles IX decreed that:

> No one may make shears or scissors . . . nor surgical instruments of metal, nor cases of falconry or any other cases furnished with astrological and geometrical instruments, unless he be master cutler, engraver and gilder.

Artisans had finally won the confidence of scientists and received many contracts to forge instruments for them. Names like Hans Lippershey and Zacharias Janssen—Dutch spectacle-makers of Middleburg—headed the industry. They had proved (with the help of principles of mathematics and geometrical optics then being developed) that lenses were not trickery, but actual aids to the direct observation of real objects. Cole and Chancellor were British names in the profession equated with high quality, as were Cock and Reeves a bit later. Instrument makers spent many years in apprenticeship and journeyman stages perfecting their skills. It became common for them to engrave their names

The mathematical instrument maker was a skilled metalsmith and optician as well. (From Phillips' Book of English Trades, *1823)*

on their products, treating them as works of art as much as manufactured goods.

No sooner had the profession reached a pinnacle of artistic quality and technological precision, however, than it quickly turned into a mass production business, earning the scorn of scientists and eroding the status of the profession as a craft. Mechanical methods were developed for marking precisely divided arcs and straight edges, and the process was perfected by 1770. As more accurate measurements were developed, the industry gradually fell into the grip of industrial production techniques. Items that had formerly been handcrafted by a single skilled artisan came to be produced by a series of less-skilled workers, each focusing on one part of the production process. One observer of Peter Dollond's workshop in England in 1769 was quite disillusioned, offering these comments:

> It must be said that the instruments in brass are turned out a dozen at a time, and that clients abroad are very much mistaken if they imagine that an astronomical instrument bearing Dollond's name must necessarily be excellent in every particular. If one of high quality is received, that is a sign that it has not been finished off by one of Dollond's workmen: often, to maintain his reputation, he has the mounting and dividing carried out by his brother-in-law, Mr. Ramsden.

By the 19th century the profession of scientific instrument maker had practically disappeared as a distinct craft. Today the thousands of instruments made for laboratory and other scientific use, as well as clocks and navigational devices, are typically designed and constructed by *technical engineers* with the aid of high-speed computers. They are then mass-produced in large factories, which usually produce many other items besides such instruments. There are many categories of skilled *laborers* and *technicians* within the industry, but as a craft, the profession has virtually died. Nonetheless, scientific instrument makers played a significant role in the history of science by helping ground it in mathematical calculation, observation, and technological innovation and processing as well as in intellectual speculation.

For related occupations in this volume, *Scientists and Technologists*, see the following:
Astronomers
Biologists
Chemists
Computer Scientists
Engineers
Physicists

For related occupations in other volumes of the series, see the following:
in *Artists and Artisans*:
Clockmakers
Glassblowers
Jewelers
Locksmiths
in *Healer* (forthcoming):
Physicians and Surgeons
in *Manufacturers and Miners* (forthcoming):
Factory Workers
Metalsmiths
in Warriors and Adventurers:
Sailors

Statisticians

Statisticians classify and analyze mathematical data—statistics—in order to draw general inferences or trends from them, as well as to establish some patterns of probability and predictability. They are involved in many professional fields and all of the physical and social sciences. For example, in agriculture they help in the process of fertilizer comparisons, to decide which are more effective and practical. In geology, they help determine the nature and extent of oil reserves. They help the insurance and actuarial fields in the establishment of appropriate premiums. Market researching, advertising, and psychological-sociological testing all demand the services of competent statisticians.

The profession had its roots in the analysis of games of chance. Gambling casinos were just becoming fashionable in the 17th century when Pierre de Fermat and Blaise Pascal, noted *mathematicians*, developed a system to predict the odds of winning games based on the concept of probability. From the early 1700's the Germans used statisticians to help gather population figures for taxation purposes. The first scientific applications of statistics were made in the 19th century by Pierre Simon Laplace and Carl Friedrich Gauss, who studied the relationship of probability and errors, respectively, to astronomy. In the 19th century Gregor Mendel made statistical studies in genetics, while others developed psychological and educational tests and measurements for statistical analysis.

In the 20th century, amid the greatest scientific and technological revolutions in history, the field has become more refined as mathematical designs have become more sophisticated. Karl Pearson worked out many mathematical bases for statistical studies. R.A. Fisher applied known methodologies to agriculture and medicine while establishing yet another method, that of *maximum likelihood*, which enhanced the statistician's skill for determining predictability. Walter A. Shewhart and Abraham Wald pioneered the application of statistics to *quality control* in manufacturing, making the field responsive to the new phenomenon of mass production. Public opinion polls and surveys also began using the skills of statisticians in attempting to assess the progress of election campaigns, marketing conditions, and vital government data in such areas as unemployment and consumer price indexes. The profession has come into particularly great demand since World War II as the age of computers has steadily evolved.

Statisticians, like mathematicians, might be thought of as operating in two realms—the *pure* and the *applied*, or the theoretical and the practical. The practical statistician plans data collection; analyzes and interprets the figures obtained from experiments, studies, and surveys; designs explicit questionnaire techniques and forms; and supervises actual polling or testing surveys. Beyond this, applied statisticians present numerical in-

Statisticians are employed in many fields, such as actuarial work in insurance. (Ralph Vasquez, Metropolitan Life Insurance Company, from Society of Actuaries)

formation through computer readouts, graphs, charts, tables, and written reports. They offer conclusions and forecasts, yet also analyze the limitations of their results in terms of reliability and usability within specific contexts. Theoretical or—as they are more commonly called—*mathematical statisticians*, generally develop the bases of statistical analysis, and work on constantly developing new and improved analytical methodologies. They research the very theories upon which probability and inference models are based, while they also test experimental designs. For the models they offer to the applied statistician, they also explicitly recommend usage within the physical and social sciences, as well as in business, industry, public opinion polling, and government.

For related occupations in this volume, *Scientist and Technologists*, see the following:

Astronomers
Biologists
Computer Scientists
Geologists
Mathematicians

For related occupations in other volumes of the series, see the following:

in *Financiers and Traders*:
Advertisers
Insurers

in *Harvesters*:
Farmers

in *Healers* (forthcoming):
Physicians and Surgeons
Psychologists and Psychiatrists

in *Leaders and Lawyers*:
Political Leaders

in *Scholars and Priests* (forthcoming):
Scholars
Teachers

in *Warriors and Adventurers*:
Gamblers and Gamesters

Suggestions for Further Reading

For further information about the occupations in this family, you may wish to consult the books below.

General

Bernal, J.D. *Science in History*, in 4 vols. Cambridge, Massachusetts: M.I.T. Press, 1965. An excellent and thorough treatment of the history of science; includes good material on the development of various professions.

Butterfield, Herbert. *The Origins of Modern Science*, revised edition. New York: Collier Books, 1962. A good overview of the progress of science in general.

Feldman, Anthony, and Peter Ford. *Scientists and Inventors*. New York: Facts On File, 1979. Sketches on the lives and works of many scientists in all periods of history and in the various disciplines.

Needham, Joseph. *Science and Civilization in China*, in 7 vols., some with multiple parts. Cambridge: At the University Press, 1954- . A wide-ranging, major work, with publication continuing at this writing.

Ronan, Colin A. *Science: Its History and Development Among the World's Cultures*. New York: Facts On File, 1982. A valuable new survey.

Astronomers

Bova, Ben. *The New Astronomers*. New York: St. Martin's, 1972. A brief look at the history of astronomy, especially in modern times.

Biologists

Taylor, Gordon Rattray. *The Science of Life: A Picture History of Biology*. New York: McGraw-Hill, 1963. A colorful and thorough account of the historical progression of the field of biology.

Cartographers

Wilford, John Noble. *The Mapmakers*. New York: Knopf, 1981. A thorough look at cartography and its practitioners; considers how cartography is related historically to the other sciences.

Chemists

Leicester, Henry M. *The Historical Background of Chemistry*. New York: Dover, 1971. A thorough treatment of the history of chemistry, with one chapter devoted to professional development within the field.

Geologists

Moore, Ruth. *The Earth We Live On: The Story of Geological Discovery*: New York: Knopf, 1963. An in-depth and biographical look at the great geologists in history and the ways in which their discoveries have shaped and nurtured the science; reads like an adventure novel.

Mathematicians and Statisticians

Kline, Morris. *Mathematics in Western Culture.* New York: Oxford University Press, 1953. Very enlightening coverage of the influence of mathematics on society, reviewing the work of the greatest mathematicians throughout history.

Kramer, Edna E. *The Nature and Growth of Modern Mathematics.* New York: Hawthorn, 1970. A thorough and readable historical survey.

Physicists

Inglis, Stuart J. *Physics: An Ebb and Flow of Ideas.* New York: John Wiley, 1970. A general history of the evolution of physics; includes readable accounts of the thinking of some of the great physicists in history, although it is quite technically elaborate regarding major ideas.

Titles in the Work Throughout History series

Artists and Artisans

Builders

Clothiers

Communicators

Financiers and Traders

Harvesters

Healers

Helpers and Aides

Leaders and Lawyers

Manufacturers and Miners

Performers and Players

Restaurateurs and Innkeepers

Scholars and Priests

Scientists and Technologists

Warriors and Adventurers

INDEX